HOW TO MAKE THE WORLD INTERESTING

The First Worldwide Book of a scientific field "Conceptology"

author: Tar Sahno

§1 Introduction to conceptology
- §1.1 The conceptologist's manual
- §1.2 Can you become a genius without talent?

§2 Creative thinking
- §2.1 The sphere in hand
- §2.2 The five items
- §2.3 Associations, synonyms, antonyms
- §2.4 The connection of everything
- §2.5 Observation
- §2.6 Neural boom
- §2.7 Random shapes
- §2.8 Write a story
- §2.9 Paradigm shift
- §2.10 At the crossroads of patterns

§3 Invention and improvement
- §3.1 Concept implementation forms

§4 Concept book creation
- §4.1 Writing up a detailed brief
- §4.2 Document design

§5 Conceptual innovations
- §5.1 By reinforcing the negativity
- §5.2 By rejecting bullying
- §5.3 By miracle effect
- §5.4 By deliberate reputation enhancement
- §5.5 By collaboration
- §5.6 By compulsive development
- §5.7 By destruction

§6 Consumer concepts
- §6.1 What do lotus and a modern umbrella have in common?
- §6.2 Walking over the mud and onto the podium

§7 Research concepts
- §7.1 Your delicious lunch through history

§8 Philosophical concepts
- §8.1 Should fate be blamed for misfortune?

§9 Economic concepts
- §9.1 Let newspapers be sold and shoes shined!

§10 Realization of potential
- §10.1 Conceptology as a profession

§1 ALL THAT SURROUNDS US USED TO BE JUST IDEAS.
INTRODUCTION TO CONCEPTOLOGY

Before we start, take a look around you. What can you see? Most likely, familiar things such as tables, chairs, other pieces of furniture, kitchen utensils, appliances, houses, shop windows, windows, cars, transport interchanges. We are used to the things which surround us. We cannot imagine our way to work without taking a bus, metro, taxi or personal car, or having an aromatic coffee in a paper cup from our favorite coffee shop. We can't imagine working without air conditioning when it's too hot outside. Many things became part of our everyday lives, and they all seem so ordinary to us. It seems that they have always been that way. To invent them, people practically didn't make any efforts, and everything organically became the way it is now. You may even feel sad when you think that so much things has been invented in the world and now is no place for something radically new, revolutionary and unusual. I assure you that it's not. And this book can prove it.

But let's go back to the items of everyday life. Everything that surrounds us used to be a fantasy and an idea, an observation, an assumption and a dream. But as soon as humanity started to have the need to satisfy its specific demand, especially if it was something tangible and material, that's when the actual material solution appeared. And this is a basic thesis you should keep in mind before starting to study conceptology.

Conceptologists have always been cool

This statement is not a theoretical exaggeration. Although the "conceptology" as scientific field didn't exist before, this doesn't mean at all that there were no conceptologists. There are many examples in history of what the first conceptologists were like. We call them revolutionary inventors, those who came up with something that turned the world of science, progress, and everything else upside down. For example, Leonardo da Vinci. He possessed not only artistic talent, but also a high level of what we call creativity today. He was significantly ahead of his time and invented something that simply could not be realized with the help of technologies and tools that existed in the 15th-16th centuries. But what Leonardo invented later became the prototype for many things we use. Why can Leonardo be called a conceptologist? Because he figured out how to solve various problems of the 15th-16th centuries for people. He didn't just invent them, but he drew them in detail. He created new real solutions of the many old concepts.

For example, let's consider a prototype of a modern diving suit. The scientist created a diving suit from leather. The helmet had glass lenses. The shoes were equipped with special metal weights. In order for a person to breathe in such a suit, the genius invented a special bell that had to be lowered into the water. Tubes went from him to the diver's helmet to allow him to breathe. Where did the need come from so that the costumes were worn by soldiers and could sink enemy ships?

The similar situation is with the first parachute prototype. In the days when da Vinci lived, there was not even any idea about flying machines. How could he invent a parachute? The idea was to create a device that would help to move smoothly through the air, "glide", and not fall. Then there was a hang glider. Leonardo was inspired by bats. He loved to observe nature and gather his ideas from there. This is how the prototype of the helicopter's propeller appeared. The genius just observed a bat when he watched maple seed wings fly from the tree. It wasn't the limit for him and this is not the limit for us.

There was also a prototype of a bicycle, a spotlight, a car. Da Vinci also came up with the first robot! It was, rather, a mechanical person. He studied human anatomy for a long time and passionately and came to the conclusion that muscles help the body to move. Then he got the idea that something similar could be created. Well, if you do not create, then, in any case, come up with and draw with details. If there were more modern technologies in Leonardo's time, then cars and helicopters would fly around the world. But just think how bold and unreal his ideas seemed at that time!

Another conceptologist is certainly Nikola Tesla. The genius was also ahead of his time and gave the world so many great ideas, and many of them are still not implemented, but who knows, they might be realized after some time... He came up with a steam electric generator, which was supposed to replace the piston steam engines. He also wanted to make this generator small, so that he could carry it in his pocket. Everyone has heard about the Tesla coil, although now it can only be found in museums. It became the prototype for a magnifying transmitter. His idea was to be able to generate electricity from a distance. Several such transmitters can provide electricity for all the world. There is a possibility that it was this idea created for the invention of the satellite Internet by Elon Musk.

A modern example of a conceptologist is Steve Jobs. He wasn't only a daring inventor, but also an excellent marketer. He knew how to satisfy the needs of mankind, and efficiently cope with various people's desires. For example, he noticed that it is important for people to stand out and create status, so he created an iPhone that almost everyone wanted. He noticed that those who work in the digital field need cool picture quality and ease of work with graphic programs, so he created the iMac. Then he came up with the MacBook, making it thin, stylish and convenient, so that people can take it with them on trips and work from anywhere in the world. Jobs set such a pace and challenged all competitors, motivating them to create cool and efficient innovations. No doubt, thanks to him there are many flagship models in the world of technology.

Should the material value of these ideas be discussed?

It might seem like becoming a conceptologist is difficult. One needs to have talent and think creatively from early childhood. But this is not enough, our brain should also be trained, and conceptology knowledge can be applied absolutely in any field. The ability to break reality into concepts helps specialists to solve their problems. It can help lawyers look differently at work cases and find loopholes that can turn the tide of the case. Conceptology knowledge will help financiers not get stuck on a task and optimize reports so that they take less time. Business leaders and owners can find new ways of development easier or points in which they can turn the business in a different profitable direction. Creative professionals can forget about what a creative crisis is. Or if you find yourself at the right time in the right place with useful information, it will help you move up the career ladder. It will help coaches and information businessmen to literally disassemble any client's problem into atoms and offer the most optimal solution, improve or radically change their techniques, etc. Conceptology is useful even in everyday life. It teaches you to question and get to the truth, even in kitchen conversations, doing it subtly and reasonably so that everyone will listen with open mouths.

Great inventors weren't the only conceptologists

Here is the story about how an absolutely ordinary person became rich thanks to his ingenuity and ability to look at things outside the box. One well-known dental company went bankrupt. And none of its top managers couldn't do anything about it. They did their best with all their efforts searching for a solution, and it got to the point that the managers simply announced the problem to all employees.

One of them said: "I know how to make your sales grow by 20%, but I will share this idea for half a million dollars. It will be a mere trifle for you, considering how much money it will bring you." At first, the management did not agree to give this employee any consideration, but soon the entire crisis management department was curious to find out what the employee had to say. After all, if an ordinary employee was able to come up with something, it means they also can and should. But there were no solutions from any senior staff.

They had to agree to the terms of the worker. When the company paid the employee off, he said only: "Make the neck of the tube wider by a couple of millimeters." It was a brilliant idea, because this way the toothpaste ran out faster, and people started buying it more often.

Here is one more favorite story about marketers. A sports shoe firm with facilities in Africa was facing a challenge. There were frequent thefts in factories. The solution was simple - to make only the left sneaker at one factory, and only the right one at the second.

A conceptologist knows how to decompose any situation into concepts, find a cause-and-effect relationship and predict how this or that concept will behave in the future. A conceptologist can also explain what to do and how to get the best possible results in the end. This is because the concept itself develops in one way or another, and everything that surrounds us already carries information from its previous forms. The conceptologist correctly processes and analyzes the information that the concept object already carries and predicts how it can behave further. This book provides all this information.

§1.1 The Conceptologist's Manual

All organized groups or people united based on various interests have their own sets of rules, so that everyone is comfortable interacting with each other, and the information they own can be disseminated correctly, especially when it comes to the merit of someone else's intellectual work. Conceptologists are no exception. Below is a list of the top basic rules that a conceptologist follows.

1. Terminology and analysis

The term "conceptology" has already been used earlier in linguistics, but exclusively as a functional-cognitive analysis of linguistic units. All studies of this linguistic field come down to the study of linguistic terminology and the origin of words. We regard conceptology as a complex science of the creation of global concepts and their development and interaction with them. Since we study many subjects and directions related to conceptology, I see no reason to leave this term exclusively in linguistics or change the names of our science due to its parallel actual use. Moreover, I believe the linguistic field of conceptual study is one of dozens of other fields.

I explain the meaning of "concept" as a clear formulation of meanings into a single integral form. When various meanings, ideas and forms function harmoniously in certain interdependent patterns, this can be considered a concept.

In addition to my assessment, one can also find interesting thoughts about this term in information sources. They assume that a concept is a collection of features necessary and sufficient to identify a fragment of the world or a part of such a fragment. It is an idea with a creative meaning. This is a stable linguistic or author's idea that has a traditional expression. This is a complex system of views on something, interconnected and forming an interconnected system, namely the system of views on phenomena - in the world, nature, society.

All of these statements are correct. Each of them makes sense based on which area was analyzed as a concept. One of the objectives of this textbook is to explain in which areas of conceptology these formulations are best suited. Every science or scientific field requires precision. The more precise the science, the more understandable it is. And this accuracy begins even with an elementary formulation of what this science is. If you open textbooks on any scientific field, you will immediately come across the fact that they will explain to you what kind of science it is, what its history is and how it is useful to society.

Biology is the science of living nature, one of the natural sciences, whose scope is living beings and their interaction with the environment.

Chemistry is one of the fundamental science of natural processes, which studies the internal composition, internal structure of matter, the laws of qualitative changes, decomposition and transformation of substances, as well as the laws of the formation of new substances as a result of qualitative changes.

History is a humanities science that studies past events.

Conceptology is a science that studies concepts, their history, the processes of creating concepts, their improvement, and the impact of concepts on society.

The first thing that guides a conceptologist in his work is the definition of terms. This is what helps to decompose the concept into its components and understand where its history begins. The main conceptologist task is to analyze the necessary semantic phenomenon, give it an expert assessment, and briefly summarize the information about this concept.

The most important is to propose the development of this concept, developing its possible innovations. But besides this, it is still necessary to systematize the information received about this concept and propose the development of the concept, developing its possible improvements (innovations).

The main subject of the conceptology study is a human being, more precisely, the search for solutions to problems in society for its more comfortable life and specialized functioning. It studies the psychological reactions of a person and society to verbal and non-verbal signals, and the reaction of various parts of the brain to the information received about how to solve a particular problem.

2. Formula of the concept

It all starts with a formula, which is simple, elegant and beautiful in its performance. It's beautiful because it is universal for all concepts, it is at the heart of everything. It is this formula that from now on will be used by any conceptologist in his work.

This is the second point that you need to write down, memorize and frame in the form of a picture on a pedestal. Let me give you an example in the form of a story to solidify this formula in your mind. Let's say we have an apartment building with 300 units, with no water supply and the tenants have nowhere to even wash their hands. From the list of possible problems, let's leave this - it is impossible to wash hands.

PROBLEM*SUB^N = PROBLEM^N

IT ALL STARTS WITH A PROBLEM. ITS MAIN FUNCTION IS THAT WHEN IT ENCOUNTERS ANY NUMBER OF SUBJECTS, IT MULTIPLIES.

PROBLEM*SUB^0 = PROBLEM^0

OF COURSE, THERE IS A POSSIBILITY THAT THE PROBLEM WILL NOT MULTIPLY WHEN MEETING SUBJECTS, IF THE SUBJECT IS PASSIVE REGARDING THAT PROBLEM.

(IDEA*J=CONCEPT)*ROBLEM^N = PROBLEM^0

BUT EVERY NEW PROBLEM GIVES A BIRTH TO "IDEA", AND HOW TO SOLVE IT. WHEN IDEA IS FORMED IN A CLEAR VISION AND PROVIDED BY ACTION "J" (FAMILIAR TO PHYSICS STUDENTS), WE GET "CONCEPT". WHEN CONCEPT IMPACTS THE PROBLEM, IT SOLVES IT AT THE ROOTS. THE PROBLEM STILL EXISTS, BUT IT DOESN'T HAVE ANY INFLUENCE.

The problem itself has no number; it is always in the singular but the scale of its influence may be increasing. If 300 families cannot wash their hands, the scale is measured in 300 apartments or even 500 people. The idea may appear - to install pipes with water in each apartment. For this, specialists bring pipes, equipment and the necessary funds. By their own efforts and action, they lay pipes into each apartment, so that water can flow from the taps of 300 apartments. The problem has been resolved and no longer has a large-scale impact. But has the problem itself gone away?

The answer is no, it has not disappeared. The problem cannot go away. It does not scale due to the presence of the instant solution provided by the concept and its implementation. But if suddenly the pipes break or water is not supplied to the house, then the problem will scale again, since the concept did not imply any breakdown or other factors. In this case, we will generate new ideas for a new problem, bring them to life and eliminate the scale of the same problem.

It turns out that the concept is not able to finally solve the problem, but can eliminate its current state, reduce or completely eliminate its scale and provide a regular fight against the problem. As for the exceptions, when the concept does not affect the subject, then within the framework of this story it is a grandfather who lives in a neighboring rural house, takes water from a well since childhood and has never used innovative plumbing solutions.

3. Searching for knowledge

You can't get out of your head what you haven't put into it. Therefore, before delving into the analysis of the concept, you need to live its history, understand and find all the forms that it can contain. Some of these innovations were still in the days of ancient people. Some, due to technological progress, time and circumstances, could appear much later.

It seems to be difficult at first, because our brain is not used to seeing all its previous forms in one object. But each time it will be easier and easier. In the era of the internet, finding information is a pleasure. To find everything that is connected with the history of the emergence of the pen and its path to the form in which we already know it today, it is enough just to wander around the expanses of the Internet.

There are many online magazines, online books, forums, articles, blogs that tell the story of different inventions. Sometimes information can be found in the most unexpected sources. For example, in fashion magazines, you can easily find not only the history of the heel, but also stories about how engineers came to the decision to create a stylish thin laptop. You should not be limited only to scientific articles, look broadly and from all sides. But do not forget to filter and check what seems unreliable to you.

It is best to organize information chronologically, from the very first mention of the subject to the present day, or if organize it geographically, depending on how the object developed or in what form it manifested itself in different parts of the world.

4. Analytical data

When working with a concept, it is important to take into account the peculiarities of the area where it is located or socio-demographic group that uses it. Due to different historical, geographical and natural reasons, everything in the world has been developing in different ways.

For example, in many countries the first money was made from what was available in the region, for example, pebbles, shells, animal skins, etc., just because in their area there was nothing else. This occurred even though the need and the idea of how to satisfy it were the same. Always consider demographic trends and perceptions of different social groups. You can search for information yourself, or you can involve an analyst.

In addition, the context of the concept you will usually analyze for a business or organization must be considered. They are also located in a certain region and are guided by their target audience. This means that, first of all, even in the history of the concept development, you should draw from local sources, and then predict what kind of development for this concept requires a local analyst.

5. Graphic design

This point is equally important. While everything that you have developed is not logically, clearly and understandable, it cannot be sold for a high price and called a valuable work of a conceptologist, especially when it comes to the presentation of your work to the customer. It is important that your work with the concept is not only analytically literate, but accessible and understandable. You are a specialist and you know a lot but the person to whom you will present your work does not understand concepts at all. It is not even this person's duty to understand the concepts, otherwise why should he need a conceptologist's assistance?

Your task is to translate everything from the conceptologist's language into one that is understandable to any person. You need to explain to a person in an accessible and easy way how to solve their request or problem. The graphic design helps this more than it might seem at first glance. Drawing and showing all of this is another task for your team. That is why in this book you come across and will come across many illustrations that complement the context of the narrative. To present their work in a really great way, conceptologists need to work with copywriters and designers. Thus, the result will be much more focused and clearly demonstrated. As I mentioned earlier, you can also involve financiers, analysts and historians to deliver the highest quality results.

6. Communication

I owe the birth of this principle and its development in the Conceptologist Guide to the Buddhist school of Falun Dafa, whose master is Li Hongzhi. In self-knowledge and self-development, Truthfulness-Compassion-Forbearance is honored there, so I'd like to ask this book's readers not to take a strict position while reading it.

Do not take my position on absolutely every issue, do not try to argue with my position at any suitable occasion, and do not throw out loud words without reading the thought to the end. I respect the opinion of each of you and ask only one thing - to listen to the thought I am trying to convey to you, and to consider the possibility of its presence in your paradigm of perception of this world.

I'd also like to ask you to be kind to everyone with whom you discuss this book, as well as to the author of the book himself, even if certain things do not fit your paradigm today. Changes happen to us every day, and in the end, you cannot be sure that some beliefs will not change over time. Perhaps what I'm suggesting to you will appeal to you in the future.

Please be patient with the knowledge gained, research, analytical data and other information provided. Perhaps many terms will not be clear to you, and you will be forced to search for their meanings. Or maybe when discussing this book with someone else, you will begin to fiercely defend a certain side. At this moment, I ask you to remember this request and show patience in relation to the entire world around you and this knowledge.

We are beings capable of observing and enriching our experience with certain facets of understanding the same situation. Many people agree that it is only in patience and calmness that deep, competent and wise observation becomes possible. This is the only way to throw away all emotions, unnecessary distractions and make a clean, unbiased and correct decision.

7. Copyright

Every self-respecting conceptologist always refers to the works of other scientists. Accordingly, we have the right to demand the same respect. Let me share a short story with you. For a long time I have been planning to publish a book on conceptology, which would be the first textbook to help people master this profession right away through real examples and practice. More than 10 years of analysis, studying information (which was very scarce), my own theories and their verification allowed me to systematize and structure my knowledge about concepts and their nature, to translate this system into an accessible and understandable language for humans. It should have been done in such a way that all methods, mechanisms and techniques can be used immediately.

One day I shared my best practices with one well-known business coach. He was delighted and realized how much these techniques will help to change his activities and improve the results for his clients. I explained to him this system's philosophy, not even the technique itself, but only its ideological nature. As a result, a month later I saw his publication in which he claimed that he himself had come up with such thoughts and spoke about this topic in front of an audience of 1,500 people. This wasn't even what upset me the most. It upsets me that he does not know the subject very deeply and gives out only superficial knowledge, which, although useful in his field, can create a fundamentally wrong idea of conceptology itself. I only agree with it if the knowledge from the book is transferred to the materials of thousands of speakers. But I ask in a humane way and legally warn that all information in the book is protected by copyright and requires a link to the author. In addition, the claim is much more solid when the speaker draws knowledge from different sources and is able to record where and what knowledge he receives.

§1.2 Can you become a genius without talent

Why is it easier for some people to achieve their goals from birth, while others need to put in a lot of effort? Why, for example, does one artist create a beautiful picture for years, while another can turn into a masterpiece in a second that was not even considered art before? Why can one person make the whole world want what he creates, while another puts his works under the table all his life. Talent and genius - how close are these concepts and why are geniuses not always talented? Let's talk about this.

Is everything predetermined by talent?

Probably, each of us has heard the phrase "If Mother Nature didn't give you talent, you can't change much"? Many people who agree on this gave up learning to draw, write poetry, compose music, etc., motivated by the fact that since Mother Nature did not reward, then it is not worth trying. Any attempt by a person to develop any abilities in himself loses all meaning immediately, as soon as he, faced with difficulties, explains their lack of giftedness. Stephen Covey, in his book The 7 Habits of Highly Effective People, calls this process a paradigm of determinism, when instead of making an effort to achieve the desired result, we blame a person or situation for not being able to achieve what we want. It can be a childhood trauma, resentment against school friends, an insult from a teacher in college or university, one phrase from an authoritative person to us, or a deep belief in things like fate or fate. It is not surprising that this weakness is used by those who are able to instill an opinion about innate abilities or hereditary shortcomings. And also to instill belief in racial difference, difference in gender, skin color or religion, in order to take possession of public consciousness and with its help to realize their political or economic interests.

For example, Aristotle argued that some individuals are naturally slaves, while others are their masters. That is, a person is not able to decide for himself who he should be, but must follow what is supposedly destined for him. And such a train of thought has long acquired a worldview value, has grown into a number of naturalistic concepts, which, based on the biological principle in a person, try to interpret the basis of personality formation. For example, eugenics (the doctrine of hereditary human health, as well as ways to improve his hereditary properties) made it possible to justify the Holocaust. Gypsies, Jews, African Americans or homosexuals have been recognized as trash that litters the pure race and must be destroyed. Based on eugenics, Hitler carried out the so-called sterilization and destroyed people, allegedly for the benefit of a pure race. Representatives of this concept received good funding for their research, the cost of which was the loss of life. The danger posed by eugenics can be seen in the documentary Psychiatry is an Industry of Death. True, here it is viewed from a completely different perspective.

Through the explanation of massive, and not isolated cases, occurs the most common manipulation of consciousness to satisfy someone's private interests. Humanity still believes that if something is common to a large number of people, then to the same extent this applies to each person individually. The important issue is that this can be easily questioned. It is the understanding that everything human in a person helps not to succumb to the influence of the naturalistic concept and to resist its champions. We can all develop any talent in ourselves, and the only difference is the amount of effort involved.

Therefore, when the goal is set, it should be achieved. If society prevents everyone to develop this way, then the structure of such a society is outdated, it does not correspond to the modern level of human self-awareness, which means it should stop existing. Everyone has a right to choose to form his social and self-sufficient essence, which will help to comprehensively reveal his personality in the future. The opinion that Mother Nature hasn't given a talent, is just an excuse for cowardice and surrender to difficulties.

Of course, I can't allow this, and the book contains detailed instructions on how to make it so that you can justly and reasonably doubt everything.

Born to be a genius... or not?

• Even Einstein parents considered him to be developmentally delayed at first; he began to speak only at the age of 4 and read at 7.

• Churchill studied poorly and was behind in his learning. His parents also did not believe that he would be able to achieve anything.

• Walt Disney was fired from a newspaper because he "lacked imagination."

• Thomas Edison's teachers called him too dumb to really learn anything.

There are many cases when an absolutely unnoticeable person with mediocre academic success and development does something outstanding. How is it that those who did poorly at school then create something great and memorable? Does it mean that education is not the key to success? Or is it not always the key to success? Ford, Jobs, Roentgen, Einstein - they were all lower-than-average students, sometimes losers and not at all from the creative sphere. But they were able to turn the world upside down. They were able to give the world something that did not exist before them and make a breakthrough for humanity in various scientific and social spheres. They came up with something that was previously out of the realm of fantasy and invention of the authors-dreamers.

We call them geniuses because they were able to creatively approach a certain issue. But have you ever thought about what if one is not born as genius, but becomes? What if anyone can be creative? I'm sure I have found the technology that will help every reader of this book generate ingenious solutions. The most important is not to drive yourself into a framework and to understand the essence of those mechanisms that help develop creative thinking in such a way that it itself offers explosive ideas. Here begins the story of the formation of our inner genius, the mysterious knowledge of his development and the first practical exercises.

What is creative thinking?

If you think that creative thinking only concerns people from the creative field, that's not true. Creativity means to creat but you can have no creative background. Creativity is characterized by two main factors:

• The ability and willingness to give birth to the fundamentally new, unusual, non-standard things, going beyond the usual idea.

• The ability to solve problems and atypical situations that arise within the system or systems.

If you simplify these two points as much as possible, then the main creativity principle is to take a familiar phenomenon or thing and use it in a way that no one has used before. Even better is if this innovation becomes necessary as a function, otherwise it will turn out to be just an interesting work of art. When we create something creative, 2 brain hemispheres work: the «left one» analyzes what we already have (facts, numbers, forms, lists, data), and the «right one» thinks about how to turn it all over into something new. Of course, I said this somewhat crudely since the brain consists of many mechanisms that work on various tasks of mental processes. Conventionally, creative thinking can be divided into two components - design and critical. They are interrelated and complementary.

• A well-developed critical part or critical thinking always gives in to all doubts and asks, "What if this is so?", "What if this is not so?", "How does it work?".

• Well-developed design thinking answers the questions "How to present something in interesting, unusual and still not exciting way?"

Even the skills of critical or design thinking separately will already be useful to any person, specialist, and employee of a large or small company. After all, the ability to look at a problem in a new way increases the chances of its positive solution. The ability to see the prospects for the development of a new idea increases the chances of a business to create a product that will solve and satisfy the needs of a huge number of people. For example, Steve Jobs, combining several technologies used separately, created the first iPhone in such a way that the whole world wanted it. Or Elon Musk turned the Tesla from just an electric car into a desirable sexy dream car.

Developing and using creative thinking is a future that came yesterday. For example, there is even the Japanese practice of kaizen, the essence of which is to continuously improve business processes, which is used by Toyota. Each employee can suggest their own method of improving work, and it doesn't matter what position they occupy at all.

Anyone can become a conceptologist, no matter his specialization at the moment. Analysts and businessmen, people from science or the creative world, those who prefer to work in an office or freelancers can also easily become conceptologists. If you are not a creative person or do not have an analytical mindset, it does not mean that you cannot create concept books and be a successful conceptologist.

Creativity is not a gift from God, but a masterful balance between two types of thinking: creative and critical. It turns out that you just have more work to do on the creative type of thinking, and not the other way around. Absolutely everyone has these two types of thinking. What matters the most is to train them right, since they're like muscles. For some reason, though, people think they must train the muscles that they can see. These create a beautiful relief of the body, a powerful and healthy frame, seductive forms - a beautiful picture. Of course, a healthy, well-trained body is beautiful. But the brain is also a peculiar set of muscles, and it requires no less attention, although outwardly they are not visible.

When a completely inconspicuous person with mediocre academic and developmental accomplishments does something outstanding. How is it that those who did poorly in school then create something great and memorable? Does it appear that education is not the key to success? Or, more accurately, is it far from always the key to success? Ford, Jobs, Röntgen, Einstein - they were all underachievers, sometimes losers and completely out of the creative sphere.

But they were able to turn the world upside down. They were able to give the world something that did not exist before them: to make breakthroughs for humanity in various scientific and social spheres.

Have you ever had the feeling that you decided that the muse had left you? Feeling like you didn't manage to come up with something interesting and original? Or you had a difficult analytical problem, and you couldn't find a solution. You began to despair, and the former enthusiasm disappeared, and somewhere in your head you said: "Okay, I have a creative crisis". What if I told you it doesn't exist? Yes! All these crises are only in our head. But if you do not delve into the essence of this phenomenon, but begin to judge the phenomena superficially, you simply cannot figure it out. However, this is definitely not about those who already hold this book in their hands. In this chapter I give a detailed guide on how to cope with the state of mind when it seems " I cannot create anything". To overcome it, you just need to perform the following suggestions. To exercise your brain - you just need to do the following exercises.

§2.1 The sphere in hand

Before starting this exercise, I will give you a short explanation. There are three types of human attention, which to one degree or another determine the level of concentration on the task, on the surrounding world and on oneself.

• The first is a small circle, a small area around a person, limited by the movements of his arms and legs. For example, when a person sits alone with his thoughts and does not notice anything around, it means that he is in a small circle.

• The second is the middle circle. It may already be a space the size of a small room. Or part of a large room. It is used when you are focused on something or in contact with other subjects.

• The third is a large circle. This is already a whole area, like a stage, an audience, a football field. The bottom line is that the boundaries of the great circle are determined by what we are able to see. For example, if we were standing in the middle of a field, then the size of the third circle would be determined by the horizon line. And everyone would see it in their own way, depending on the visual acuity and the ability to see the most distant details.

To easily switch and become aware of yourself in these circles of attention, the following exercise will help:

1. Sit comfortably and extend your palm in front of you.

2. Focus on the palm so that its boundaries begin to blur.

3. Imagine that a small sphere forms on your palm.

4. Its outline is faint, but already noticeable. You begin to see the sphere in your palm more and more clearly.

5. What color is it? Answer the question and keep looking at this sphere, this color.

6. Add texture to the entire area of the sphere, it can be anything: a soccer ball, a smooth surface of water, a brick wall, rough sandpaper. What do you have?

7. We keep the sphere in our hand, and now it grows and grows.

8. It grows and no longer fits in the hand. We have to hold it with both hands.

9. The sphere grows, and now a certain mechanism appears in it, something very fast and energetic.

10. The sphere tries to break free and slip away. It is as if she is trying to jump out of your hands, and you are putting all the efforts of your imagination in order to leave her in the field of your influence.

11. Now it is already spinning at a high speed with which you can only imagine. She is ready to break free, but you deftly keep her in the middle of your zone of influence. We count from 10 to 1. And after saying one — "bang"!

12. The sphere exploded, and the entire room was completely covered in the color and texture of the sphere.

13. And then it all slowly starts to slide off the walls and eventually disappears.

14. The room returned to its previous state.

Why do we need this exercise?

It helps "start the gears in the brain", using both rationality and creativity. The part of the brain that is responsible for creative thinking and that that is responsible for critical thinking works. It invigorates and stimulates neural connections to get ready for work. This exercise is recommended to be done daily, better before work. But do not do it at night - such active dreams may not give you rest.

§2.2 Five items

Before we start, here is a bit of history about the game of mahjong, the prototype of this exercise. Mahjong is a great memory trainer. The point of the game is to free the tiles with the images that are under the others. This can be done only by memorizing the same images and making the correct moves

There are different mahjong variations, but they all boil down to one thing:

• Those game pieces that are under the others remain stationary.

• You can only remove 2 identical tokens.

• If you make the wrong moves, you end up not being able to remove all the chips and win.

This game is great for both adults and kids. There are many variations of mahjong created for computers or mobiles. If you want to train your memory and attention - download mahjong to your phone and while away the time in traffic jams or on the way to work profitably. By the way, versions for a computer or smartphone are usually more dynamic and colorful, without taking up as much space as a box of cards or tokens.

It's time for exercise:

1. Pick 5 random items near you. For example: a pen, headphones, a ruler, an old cell phone, a medical mask.

2. Place them in front of you in any order and take a photo (it will be easier later).

3. Memorize every detail, even the smallest one. Memorize where the pen lies, its relation to other objects, whether the inscription is visible. Is the mobile phone on or off? Do the airpods have their pads up or down? Every detail. You should take a moment for this.

4. Now shuffle all the items.

5. Try to restore everything from memory to the details. All the locations of things and everything is as it was before you shuffled the items.

6. Compare the results with the photo.

Exercise benefits

It trains your memory and develops new neural connections that will help you memorize things perfectly and quickly. Analyze what methods you use to remember the order of things. How does your memory give you images one by one? Are you creating associations or visualizing a picture? The exercise will get better and better each time. When it becomes very easy, increase the number of items. I also recommend practicing in pairs. This makes the task more difficult. The "presenter" will need to arrange the objects "with a trick" - to think about how to make the "player" remember as few details as possible. And the player, on the contrary, needs to remember as many details as possible. Try it in pairs. You will certainly be surprised at what miracles our brains are capable of. **Important:** this exercise should be constantly complicated. Once you feel that you can handle 5 subjects easily, add another one. Gradually increase the number to twenty or more. It will be quite difficult, but your memory will work better and better.

§2.3 Associations, synonyms, antonyms

>This is a group exercise. It's perfect for a group of up to 6 people.

1. Each participant is assigned a number, or a specific queue is selected - this is the order in which each will respond. But this is only for convenience, to avoid confusion when the participants are quite active.

2. First stage. The first participant says any word, the second - an association to it, the third - an association to the word of the second participant, the fourth - an association to the word of the third participant, and so on. For example: swallow - flight - plane - wings - KFC - colonel. But be careful. You don't need to take into account the entire associative array. You should concentrate only on the word that the last person said.

3. Second stage. The first participant says any word, the second is a synonym for this word, the third is a synonym for the word of the second participant, the third is a synonym for the word of the third participant, and so on. For example: wheelbarrow - vehicle - transport - car - mechanism — clock.

4. Third stage. The first participant says the word, the second says the antonym, the third - the antonym to the word of the second participant, the fourth - the antonym to the word of the third participant, and so on. For example: good - evil - happiness - grief - success - defeat.

Note: It is very important to speak exactly on the last word spoken by the previous participant, and not on the first word of the first participant or the entire list of words before that. You can do this exercise yourself using a notebook, notebook, piece of paper or a computer/phone monitor. The main thing is to save the results so that you can analyze them later.

Why do we need this exercise?

As you may have guessed, you need this exercise to pump out-of-the-box thinking and generation of non-standard solutions. This exercise will help you create new ideas and get out of the "blank sheet" state, when before completing a creative task it seems that there is absolutely nothing in your head, all your ideas seem boring and hackneyed to you.

Do not be afraid to have some ridiculous ideas. Often, they then suggest fantastic thoughts. How to put it into practice? For example, you want to write a cool Instagram post, a news story for a client site, or an article for your blog. Short and one sentence. But nothing comes into my head. We take the word head and legs. Now we process them in this way. Head - thoughts - clean hair - freshness. Legs - old man - tiredness - forgetfulness. What can we combine? "Fresh thoughts do not come into a tired head."

§2.4 The connection of everything

The principle is the same as in the previous exercise.

   1. We take two absolutely unrelated words. It is desirable that they be as far as possible categorically from each other.

   2. The participant must come up with a synonym for each word.

   3. Further, from the resulting two words, the rest of the participants in turn must come up with one phrase / word.

For example, we take the words: "deer" and "sex"
Deer - horns
Sex - breasts
Collocation: horned chest.

Airplane - flight
Surfing - board
Collocation: flying board

USA - White House
Pie - food
Collocation: edible white house

Why do we need this exercise?

This exercise is a continuation of the previous one. It helps to get out of the creative spasm and find ideas where they seemed to be impossible. This exercise helps to see the connection of everything with everything else and to be able to connect what seems to be disconnected. As experience shows, this is how cool and non-standard ideas arise and the way how our brain works. We often don't fix databases that are firmly entrenched in two different neurons until we hear or think about their combination. They are incoherent in our brain and do not make sense until a certain point. This exercise does not just form new neural connections between concepts that are close to each other, but connects together those neurons that, perhaps, would never have been found if we had not made an effort. Many inventions that we are successfully using now appeared precisely because someone came up with the idea of connecting something that, at first glance, is unrealistic to connect.

§2.5 Observation

Choose any living creature that will be the subject of your observation. It can be the person whom you met on the way to work, an animal that was seen on a walk, or a child rolling on a swing.

2. Observe them for some time.

3. Describe what you observed. What do they like? What do they do? Why do they do it that way? What do they think at this time?

1. Now describe your subject of observation in a small sketch. You have a minute. You can, of course, write longer.

For example, let's say you describe a man you saw on an airplane when you were on a business trip. You noticed that he is looking around and is clearly dissatisfied with something. You have taken a good look at his clothes and carry-on luggage. First you describe him: this is a man, he is sitting on the plane and looking at people. He's definitely over thirty and a little bald. He is wearing a red T-shirt and simple jeans. He wears sandals with socks of not the most fashionable type. But he, apparently, loves comfort so that the belts do not rub his legs.

He gets nervous and gets the attention of the flight attendants. He looks down on the passengers, thinking, "I shouldn't be here. My place is in the business class." After some time, the man fell asleep. This was the moment the plane was taking off.

And now a short sketch: "A short bald man was sitting on the plane and was nervous. He constantly looked around and could not sit down in his chair. A stewardess came up and they began to discuss something loudly. He waved his hands, and she, keeping calm, offered him something. As a result, the man waved his hand and calmed down. The stewardess left. He continued to look down on everyone and get angry. Every now and then he opened and checked something in his bag. It's strange, as soon as the plane took off, the man calmed down and fell asleep." It is important that your sketch is not complicated, but logical. **So that later this story could be shown.**

Why do we need this exercise?

This exercise helps you notice minor details, build cause and effect relationships, draw conclusions, and apply methods of deduction and induction. Yes, Sherlock Holmes probably practiced in this way too, because his investigations are based on the method of deduction.

By applying this exercise, you will be able to notice what is hidden from the view of inattentive people. You will be able to put yourself in the shoes of other people and develop different stories in your head on their behalf.

§2.6 Neural boom

1. The facilitator conducts the exercise.

2. The optimal number of participants is 6.

3. Each participant gets on a specific question in a certain order, to which the person must answer.

4. The order of the participants' answers is discussed before the start.

5. You need to answer the questions in such a way that you end up with a coherent story.

6. Until the very end, the participants do not know each other's answers.
7. They only recognize them in the process of dubbing the final story.

8. It looks like this.

The first participant writes in one sentence the answer to the question: "Who is this?" The second answers the question: "What kind is he/she?". The third: "What is location of the story?" The fourth answers the question: "What did you do then?" Fifth: "What is the reason of the previous action?" And the sixth "What happened in the end?" When all the participants are ready with their proposals, the moderator reads the questions in order - the participants voice what they wrote.

For example:
   1. **Who** — a strange man
   2. **What kind** — painted with bright colors
   3. **Where** — on the top floor of a skyscraper
   4. **What is the action** — he is pining for his old first love in school love
   5. **What is the reason** — to save the world from terrorism
   6. **What happened in the end** — making a film based on real history

Another example:
   Who — a middle-aged woman
   What kind — with a red face
   Where — in Paris
   What is the action — recalls Pablo Escobar
   What is the reason — to run away from herself
   What happened in the end — she goes into the sunset to the sound of music

And one more example:
   Who — an adult gnome
   What kind — bald and tall
   Where — in the mall
   What is the action — runs away from the wind
   What is the reason — to seem better than he is
   What happened in the end — good wins

Why is this exercise needed?

This exercise also develops creativity. But here the fact is that our task is not just to answer questions, but to write our answer into an already existing history.

Adapt it and adjust it to the logic of the story. It is important that everything sounds solid and organic, but the answer itself does not change. **For example**, if this is a bald clown who sits in the toilet, and you have it written **"She opens a bottle of champagne in honor of her birthday,"** then there are two important things to consider:

1. that it is not SHE, but HE

2. and it is not worth writing in too much detail. Since "in honor of the birthday" can spoil the answer of the next participant.

The exercise does not **just stimulate creativity**, it teaches you to fit your ideas into the existing context and predict how they may be interpreted in the future.

§2.7 Random shapes

1. Take a piece of paper or any surface to paint on.

2. Draw any abstraction on it. You can just randomly move the pen over the paper.

3. Fix the result and begin to fantasize, part of which character, situation, which element or location this chaos could be.

4. Then, from this abstractness, you need to create an understandable drawing. Abstract to concrete. The main thing is to be able to understand its essence.

Why do we need this exercise?

This exercise shows us how to create something meaningful out of nonsense. It helps to develop imagination and draw out of it those images to which we are accustomed, those images that seem familiar and understandable to us. It is not at all necessary to be an artist in order to turn an abstract blob into an understandable drawing.

Here you just need to use the part of the brain responsible for creative thinking and let your imagination run wild. Remember, as a child, we often looked at the clouds and tried to see in them the outlines of familiar figures: animals, fairy-tale characters, plants, etc. The principle is similar here as well.

Complete the blots in the row in such a way that you get understandable images:

§2.8 Write a story

1. We name or write out three words that come to mind.

2. They don't have to be related to each other.

3. Based on these words, we are writing history.

4. Words can be swapped.

5. The story should be short, about only one sentence per word, but meaningful.

6. All sentences must be logically connected.

For example, words: basin, elephant, flower

The Story:

"In the city zoo, a girl was constantly washing an elephant using a basin. It was convenient for her, because the water did not splash and wet everything around, except for the elephant. And the elephant liked it so much that once, after bathing, he even gave the girl a flower."

Why do we need this exercise?

This exercise helps to fire our imaginations and put our imaginations on the right track. Building logical connections between seemingly incoherent words, we understand that, in fact, ideas around us and our brain are seething with different sentences. It is only important to collect them and send them where we need to.

This exercise also helps to cope with "no idea syndrome".

Even when you get ridiculous results or funny stories, it still enhances your creativity. Do not limit yourself to just three words. Increase the difficulty gradually, start with three, then five, then ten or more. It will be harder, but more fun and productive. This exercise summarizes your success in developing creative thinking. If you have chosen a couple of exercises for the day (there must be a Sphere among them), then the "exercise with stories" will be a good result of exercises for a set or for a period of days. It will show how successful your thinking training has been.

§2.9 Paradigm shift

1. First you need to choose a situation where there are several parties involved.

2. Next, we blame the first party and justify the second. State a few arguments.

3. We blame the other side and justify the first. We also present the arguments from this perspective.

4. We blame both sides. And we give arguments why both are to blame.

5. We justify both sides and say why no one is innocent.

Example: "Mom made mushroom soup. The family had lunch, everyone was poisoned and ended up in intensive care unit. "

Why is mom to blame? Here are the reasons:

1. She didn't check the mushrooms and didn't consult with mushroom pickers.

2. She cooked without adhering to cooking standards, and arsenic or other household poison got into the food.

3. She got distracted and picked the wrong mushrooms in the forest.

4. She doesn't know much about mushrooms, and she cooked soup from some unknown source.

Why is the family to blame?

1. They themselves asked mother to go pick mushrooms and cook soup for them, although they knew that she was not an expert in this.

2. They were lazy, they didn't want to cook themselves.

3. They themselves chose to eat soup, although there was other food at home.

4. They slipped a toadstool to mom to SET IT UP, but they forgot about it.

Why is everyone to blame?

Mom didn't check the mushrooms she was cooking, and the family asked her to cook mushroom soup, knowing that Mom was not an expert.

Why is everyone innocent?

Mom cooked food because she was worried about what to feed the family. The family ate because they always trusted their mother, and they could not even imagine that such a situation could happen.

Why is this exercise needed?

It is useful not only for those who see themselves in working with concepts. This exercise is useful in everyday life as well. The basic rule of critical thinking is to doubt everything and always ask yourself two questions: "What if this is so?" and "What if this is not so?" This will help you get to the bottom of the truth yourself, questioning everything.

You will not allow yourself to be misled and manipulated by your consciousness. For example, when you watch the news, you always see the material that a lot of people have already worked on: editors, screenwriters, PR people, and others. Therefore, you will see exactly what they want to show you. But if you always doubt, then over time you will begin to notice all the inconsistencies in the presentation of the material, and you will probably know what could have really happened.

§2.10 At the crossroads of patterns

1. Let's take some kind of verbal semantic unit.

2. It can be one word, phrase, or sentence.

3. And we are trying to visualize it in an unusual way.

4. In order not to think about the phrase for a long time, you can take a well-known proverb or come up with an interesting expression.

For example:

We have an art gallery, and you need to show its peculiarity, through the proverb "Whose cow would moo?"

How can this be done?

"A billboard is in front of us with an art gallery wall painted. There are two paintings on the wall and a cow is painted on each. There are drawings from different artists and in different styles. Below the paintings are the signature "Whose cow would moo for you?"

Why do we need this exercise?

This is an exercise for the brain to learn to fill everything with meaning. Such practices are used by advertising agencies in brainstorms when they need to come up with a logo, philosophy, or brand positioning so as to fit it into a short, understandable phrase. In addition, to make it original, memorable, and understandable for users. Filling ordinary forms with meaning is the task of creative thinking.

Summary of all exercises. By using these exercises and practicing every day, you will definitely never have what is often referred to as a "creative crisis." You can always find fresh ideas and solutions even to those problems that still seem insurmountable to you.

One of the biggest secrets is that we are not by nature predators or herbivores. First of all, we are observers! The ability to observe, draw conclusions and influence the world around us is exactly what made us intelligent and drives our evolution. By far the most important trait of a world-changing observer is courage. So here I am, your faithful servant of humanity, the simplest person who invented a new scientific direction "Conceptology". I want to convey this courage and confidence in my abilities to each of you. You are capable of becoming a conceptologist and present brilliant concepts to this world.

Although by and large it is difficult to say that we are actually capable of creating something really new. After all, the very concept of "invention" presupposes that we have taken something that exists and presented it in a completely new form. On the evening of Friday, November 8, 1895, when the assistants of physics professor Wilhelm Konrad Röntgen had already gone home, he continued to work. He again turned on the current in the cathode tube, which was closed on all sides with dense black cardboard. Suddenly, a paper screen lying nearby, completely covered with a layer of crystals of platinum-cyanide barium, began to glow with a greenish color. The scientist turned off the current - the crystals stopped glowing. When the voltage was applied to the cathode tube again, the luminescence in the crystals, which had nothing to do with the device, resumed. As a result of further research, the scientist came to the conclusion that an unknown radiation emanates from the tube, which he later called X-rays. This is how the world's first X-ray was created!

But there were prerequisites before ...

The point of this idea is that the cathode tube itself was invented earlier by other scientists, and platinum-cyanide barium was discovered even earlier. Based on Röntgen's work, other great discoveries were made by scientists Antoine Henri Becquerel or Pierre and Marie Curie. But there is a question: did they invent something new or is it all a long series of improvements by other inventors. After all, it turns out that you can generally dig deeper and start this series with the inventions of the first tools at hand, then, through the prism of time, go through the discovery of the first metals, the invention of glass, Mendeleev's discovery of his table of elements, and so on. It turns out that the very concept of "invention" is not to create something new, but to improve something old or to notice something unknown among the world of known things.

Why do we accept new inventions or improvements only at a specific time?

Ever since childhood, we feel what is interesting to us and what is not. This takes into account color, shape, volume, and pitch. The modern child has a completely different value of time, and usually instantly reacts to what he needs and what he does not need.

In fact, all our doubts come to us with age, when we already have life experience, and as a consequence, there is a fear of making a mistake. When it comes time to choose whether to buy some kind of product, whether to settle in a new apartment, whether to pay for the subscription of any resource, we begin to analyze for a long time and doubt any of our decisions. But inside each of us there is always such a small child who knows very clearly internally when they want to "sell" a thing he does not need, and when in front of him is something new, very necessary, very important and valuable. If this child falls under the influence of trends and the popularity of brands, is led by the beauty of the wrapper, the uniqueness of the site, the usability of the application or good reviews, then later it may still remain dissatisfied, and in addition, offended. This may be due to a lack of immediate benefit and need for the item or service. The adult bought it, but the child is not happy. After all, the short-term joy of attention, comfort, self-esteem and much more passed almost immediately after purchasing the product. And the inner child feels cheated. With such an interaction, the connection between the owner of the concept and his client is irretrievably lost.

Pretty wrap isn't that important

Your concept does not need a beautiful design, beautiful presentation or experienced promotion as much as your brand, product or company needs a sense in the first place! You should not only be popular or important, but also necessary, essential for human-beings. Let's ask ourselves the question: "What is more important to us is to earn a lot of money, to look more beautiful than others, to be more fashionable than others, or, all the same, to be happy?" And we definitely come to the answer that our inner child first of all wants to be happy.

Our psychological and physical condition directly depends on this. All these conventions are just repelled by what makes us happy. All priority things for us initially start from this reason and seem to us the only right decision on the way to our happiness! Thus, introducing innovations into any concept, we simply have to take into account not some kind of marketing ploy or economic deception, but take care of the state of happiness of each end user. Take care of them the way we take care of our own children. We must anticipate and use their needs in order to meet them and to encourage any dream. And do not be afraid - the enrichment will follow by itself. But in a completely different context.

What if the script can be written?

Just imagine if such a direction of various concepts existed in politics! Each new presidential term and new election would describe in detail the way to achieve their goals. Each party and candidate would not care about how to psychologically influence the masses, but how to really propose a scenario for the implementation of these promises. And thus, each new election clearly analyzed all aspects of the problems of society and proposed realistic terms and solutions to such problems.

Such a pursuit would turn into real intellectual races with the goal of being the most productive and realistic in their plans.

Do you remember the excitement created by each new iPod, iPad, iPhone, Mac, released by Steve Jobs? We were waiting for the release of a new device, being clearly confident that we will be presented with a new miracle - the invention of the future! What happened next? The seventh iPhone comes out. The eighth iPhone is practically no different from the seventh - a couple of functions in the camera, the processor is imperceptibly faster.

The tenth iPhone comes out with a number of new innovative additions, but not at all larger than people already expected. Just imagine if the seventh iPhone came out in the form of the tenth! There are no buttons now and your fingerprint is not even needed, now there are two or even three cameras, portrait mode and many other functions, Apple Pay, face recognition, a convenient screen - that is, they would offer a miracle that people could not even at that moment to imagine. As a result, we would definitely want to purchase this product, but at the same time we would be satisfied. We will soon deal with such things further.

§3.1 Concept implementation forms

An idea is not a concept yet, and it is essential for those who are going to seriously develop in our profession, transfer this knowledge to others, create concept books and make money on it. Nobody will pay you because you tell some interesting, intriguing idea. No one will become seriously interested in those who voice the solution to a problem or question without clearly supporting it.

People want to see something tangible. Therefore, it does not matter what form your concept has acquired - it must be formalized. In this book, we will learn with you not only to analyze and disassemble concepts, but we will also be able to design them so that, as a result, your client will receive both an analysis of what is now and a guide to action for the future.

Starting to work with a concept, we immediately ask ourselves: "How to understand that this is a concept?" In addition to referring to a formula that each of you probably remembered, we also understand that concepts are impossible to touch physically.

Let's say our grandmother told us a fairy tale that her grandmother told her, and so on. But this tale has not been written anywhere. Moreover, the fairy tale itself is a ready-made concept that solves several problems at once. This is the problem of children's sleep, spending time interestingly, explaining some life observations, and trying to cheer up or scare. If something solves a problem, it can be called a concept. But again, we return to the fact that the concept of a fairy tale is impossible to touch.

And then what to do about it? How to develop or improve such a concept? I'm even afraid to imagine how many thoughts you'll have if I ask you to pick apart concepts such as prayer, song, theory, or discussion. But don't bother just yet - there is a lot of practical and theoretical curriculum ahead of you to help you understand these conceptual processes smoothly and comfortably. There are two forms of concepts.

Actual manifestation

They have a concrete form. They can be touched, are easy to describe, and their functions are usually elementary. Before an idea is formalized into an actual concept, it is described in detail, down to how it solves a specific problem or query. Then its implementation is invented, and the most optimal and effective variant is brought to life.

This is how all inventions appear: from the wheel to the shuttle. All the everyday objects around you are actual ready-made concepts. And if you dig deeper, you can look at the same object from entirely different angles, spinning other concepts into it: economic, industrial, marketing, etc.

After all, absolutely any of them satisfies your needs. Not necessarily a basic need. It could be a creative need - for example, a notepad, graphics tablet or a photoshop program.

There's also an aesthetic need - a beautiful but utterly impractical tablecloth in the kitchen or uncomfortable designer shoes that make you feel like the king of the world, but you can only walk in them for 10 minutes. When we further consider the properties of the concept, which properties satisfies the purpose, what led to these or those improvements, we can indeed predict the ways of its development and say exactly which of them will be the most successful and will lead to success.

When working on a concept book, we can combine different factual and oral concepts together. We can do this using our earlier exercises, which I hope you will practice quite often. The more time you spend on it, the more efficient and flexible your brain becomes. Let's say, using the oral concept of freedom and fully analyzing it, we can combine it with the actual concepts of an airplane, a music concert, a political candidate, and many others.

Oral manifestation

They still solve some problem or satisfy a need, but they have no tangible form. They are denounced in the format of words: legends, fairy tales, legends, treatises, philosophical trends, literary works, and the like.

For example, people often ask themselves different questions about why some are doing well from birth, while others get it all through hard work. Previously, different inequalities, like everything incomprehensible, were explained by super powers. Some people could do everything, while others have been restricted.

So, they began to talk about true freedom, which influenced and chose how to live. So many theories have appeared about what freedom is and why some have freedom and others do not. Freedom is an example of an oral concept. You have never seen an object that would be freedom, have you? But on the other hand, there are many books, paintings and even musical works that show this freedom from different sides. The search for answers to the question of what real freedom is - this is also the satisfaction of some needs, all from the same pyramid. But you yourself can already tell which ones.

The oral form of the concept can be realized. Take a book for example. Is the book a factual concept or an oral one? Pause and don't read on until you answer this question. I'd be very grateful if you take some time to think about it or discuss it with someone right now. This practice will be very useful in all the topics that we discuss here.

In order for these lines to appear in the book, we work with a large team of scientists, consultants, historians, artists, copywriters and managers. We meet to define various terms, analyze concepts and try to parse as deeply as possible each thought that we further describe here for you, my dear readers and conceptologists.

So, I hope you've already thought about what the concept of the book is. And this is the answer. What can we write in the book? Fairy tale, legend, myth, theory, teaching, lessons and much more. In fact, these are all oral concepts. But the book itself is the carrier of this information, not information.

A book is a set of blank pages on which information is written in ink in some language or in the form of numbers and signs. She is a participant in concepts such as a USB medium, a hard drive or a huge server. The book itself is an actual concept, and keeping information for a certain time is its manifestation. But the information that is written or drawn there, be it a report or drawings, are already oral concepts.

But if there are images that absolutely independently reflect a solution to some problem in a creative form, if on each page of the book there is a photograph of a shoe, a picture of a famous artist or a fingerprint, then it turns out that the book is capable of storing other factual concepts as well. After all, all of the above are also actual concepts.

What forms do the following concepts refer to?

Song:

Smartphone:

Refrigerator:

Respect:

Bottle:

Car:

What forms do the following concepts refer to?

Stadium:

_____

Chess:

_____

Football:

_____

Profession:

_____

Big Boom:

_____

Collaboration:

_____

§4 CONCEPT BOOK CREATION

It goes without saying that developing creative thinking and doing research in the field of conceptology is very exciting. This path will be useful for any person-even someone who is not going to engage in conceptology as his profession. But we live in a capitalist world, which means we need to make money. And preferably the more the better. A conceptologist is not a profession for the poor ones, but an opportunity to earn very decent money with the help of your own unique view of things.

The product that the conceptologist creates is a concept book. This is a collection of concepts that the client uses in his project, as well as the analysis of these concepts through a certain schematic analysis and recommendations. This is the basis for all departments involved in the project. Such a database of concepts presented in the project, as well as the number of practical recommendations, will save the project manager a lot of money on small and large analytical work.

All the knowledge you have received before will certainly help you in creating such a product. But so far there is no structure or specific methodology in them that would help all of you not to reinvent the wheel every time a client calls on it. That's why I have already invented such a systematic approach for you. Moreover, during the following examples and explanations for filling the brief, I came up with a non-existent client and his fictional products in order to give the most understandable examples in this business. Any coincidences with real companies and their products are completely random and unintentional.

To implement or not?

Let's see what a concept book is, considering related areas. For example, sales managers often use sales books in their work. These are documents with as much information as possible to help them sell the product better. These are the client's pains, his needs, a portrait of the target audience, competitor analysis, and so on. Brand makers use brand books, a kind of presentation, which contains all the information about the visual and ideological content of the brand. These books include colors and their meaning, as well as the rationale why they are this way. There will be images used in brand identity and reasoning why they are. The various archetypes involved, logos, icons, pictures - literally what will constitute the visual core of the future brand and all the accompanying ideas. Moreover, the analysis of the meanings that need to be conveyed to the consumer through the visual component.

But this still has nothing to do with the analysis of the concepts that the company uses. And here, either the above-mentioned experts should start using such an analysis, or we should provide it as a separate service. Both options are correct, we can use the concept book as a separate concept analytics service, but also implement it in the brand book or sales book. At the same time, it in itself is also a useful component for other departments working with external and internal processes of the company. I think you understand, these are just a couple of examples. In the world of such different directions and departments of the company's operational processes, where it is possible to introduce a concept book, there are much more.

Why is this the most useful tool in business?

The concept book is not just a beautifully designed collection of data but a large-scale tool that will then remain with the client forever. Literally everything can be done based on rebranding, crisis management, business scaling, expansion, and implementation in different areas, etc.

It is possible because concepts contain a variety of facets such as visual, economic, analytical, and creative. If, for example, we compare a conceptologist and a brand manager, then the services of the former will significantly benefit all of the people concerned. Because he not only understands the visual component and ideological nature of the brand, but also understands the entire life history of the concepts used, from the moment of their inception to the moment in which they function now. But brand-making complements the concept in the same way as vice versa. It's like in a Lineage game - by combining the skills of a concept scientist with the skills of other professions, you strengthen both professions and make services doubly useful and multifaceted.

A concept scientist with a branding background can analyze it from how trendy a concept is and how viable it can be in interacting with current consumer preferences, but also predict dozens of future prospects for the concepts being used. As an expert in the field of concepts and in the field of creating new brands, such a person can predict many more options for the development of a concept and advise the one that will be the most optimal both in terms of brand presentation and in relation to its most hidden pain points.

I am definitely in favor of using conceptology especially at first as a related profession and using concept books as an addition to your already existing services and skills. Actually, that's how I started. A concept is a collective concept, and in the process of working with this data, it is impossible to simply ignore other departments of the company. This is what makes the concept book kind of a unique business tool for achieving goals. Next, we will figure out what it consists of.

§4.1 Points of tangency: writing up a point brief

Frankly speaking, I did not come to what I now call a "point brief", as well as to the concept book, which we will talk about further, too. Rather, we are presenting this in the form in which it is now received by my clients, using the technique from this book. Initially, I discovered that there are concepts around us, then I realized that these concepts are improving, and called them "conceptual innovations."

I could not even imagine that this would turn into a whole hidden scientific direction, which is essential for humanity and our future functioning. When I realized that there are different concepts, I researched the processes, and how exactly they are improved. Questions popped up in my head regarding the natural course of improving any concept, both artificially improved and intentionally stimulated improvements.

While creating visual design and IT products for various companies, I later came to the conclusion that I would give much more analytical information during the preliminary consultation. So why shouldn't this information, which is not yet the most creative one, be presented properly? After all, it is one thing to just talk about important points, and it is completely another thing to record the information and give the client the maximum.

In addition, this approach helped me to see more clearly all the outgoing data for further creativity, working with products and improving its concepts. It was from this that the main points of the point brief were born, which relate not only to innovations. However, let's start by looking at different examples of briefs.

A brief is a document, a short written form of a conciliatory order between the parties planning to cooperate, *in which the main parameters of the future software, graphic, media or any other project are prescribed.*

A brief questionnaire is a technical task for one of the parties *to request more detailed, preliminary information about the transaction.*

A creative brief is a document that helps a company to correctly set the task *of developing an advertising message or image of a product, creating an advertising product (video, poster, etc.).*

An expert brief (a brief on the creation and introduction of a new trademark) - its questionnaire part, offered to the client, *is partly a tool of marketing research that precedes branding - the creation and maintenance of a trademark.*

A design brief *is a short-written document that defines the desired outcome of a design project.*

In our case, the brief is used to fix the points of analysis. When we have enough collected information about some concept, we start analyzing it. In the brief, you can make tips for yourself, which will then be used when communicating with the client. For example, a problem, which we will consider its impact on society and options for solutions in the past. In order to help you as much as possible to understand how to draw up a brief in a concept book, we will be immediately giving examples, so for analysis I decided to take something simple, something that people use all the time.

The examples will be something that you probably know a lot about and with which you yourself can work, for example the concept of drinking coffee.

From this point on, the book becomes even more interesting, because now you will combine theory and practice. You can safely grab a notebook and a bunch of pens to practice writing your own brief. You can also use your computer to immediately structure it and get used to typing the summary of your thoughts into a finished document. I also want to advise you further to go through only one or two paragraphs a day. If you feel overwhelmed, you can take breaks between paragraphs.

If you are using this in your school curriculum and mixing this data with other subjects, then it is best to absorb one or two paragraphs a day, twice a week. This is really important, because the goal of learning is to ASSEMBLE information.

Point One - A Brief History

Like in Yuval Noah Harari's book, I always tell the client a short story, though without focusing on the story itself, but rather on its key points. A hundred years ago there was such a problem, then it was solved in this way, but the problem has remained unresolved to this day. Or it was solved in different ways that no longer work, and the existing problems of this concept can only be solved by new methods. At first glance, now it can be solved this way and that, but let's delve into the very essence of this concept and try to look at how contemporaries looked at it and how ancestors looked at it.

It is also important to see the pros and cons of the project, describe them and think about how you can use them in the future. In fact, the study of these problems is the most important, because this is what the whole concept book, all innovations and solutions are based on. If there were no problem, no one would have come to us for a solution as conceptologists.

At the stage of a short description, you need to analyze all the ins and outs, re-read the history, information about the concept, understand what you will use in the future, whether the concept can become a trend or it needs to be collaborated with a trend for viability, or it has died altogether and you need to figure out how to revive it, or resurrect it into something else. Only critical thinking works here, so we turn it on and critically analyze the collected information.

A short description of the project.

People traveled the world, discovered a new plant, tasted it first, then accidentally fried it, and so that the product would not deteriorate at all, ground it into powder. This was analogous to the processing of spices, which was already

A short description of the project.

understood at that time. Then someone accidentally poured water over the powder and got a very aromatic drink, which later grew into the culture of coffee drinking.

Next, we analyze how everything happened: how they drank in different countries, how they imported to which country, and where they first began to produce coffee on an industrial scale.Now you need to find living points that this concept or product touches.

*They can be divided into concept pros and cons.
Pros of coffee drinking:
- There is a whole culture of coffee drinking
- Coffee helps to cheer people up
- Coffee is now free of bugs and debris when it is transported.
- There are many different ways to make coffee.
- Everyone can choose according to their taste
- The world has come up with cozy coffee shops that are so cool and inviting to go on dates, tune in for the day, or hold business meetings.

Cons of coffee drinking:
- A lot of coffee is bad for your health.
- Coffee abuse makes teeth yellow
- Coffee is addictive
- Coffee invigorates only for a short time.
- Many people died during the transportation and extraction of coffee

Write a short description of another concept for hands-on learning.

_____

_____

_____

_____

_____

_____

What are the advantages of your concept?

_____

_____

_____

_____

What are the cons of the concept?

_____

_____

_____

_____

Point two - the author's analysis

Author's analysis

The process of making kopi-luwak coffee beans consists in the fact that musangi (Paradoxurus hermaphroditus, an animal of the civet family) eat ripe fruits of the coffee tree (coffee cherries), digest the pulp surrounding the coffee beans and defecate coffee beans, which people collect, wash and dry in the sun.

For information on the "kopi luwak" coffee variety, try to research this topic and learn what a person should have in his head to try it for the first time. Probably, at some point, ordinary things get boring, and you want something new. Or the animal accidentally tasted a whole batch, and it was necessary to be creative in solving this problem. At first this idea seems ridiculous, but then in the process of brainstorming it is she who can give compelling ideas for an advertising campaign, logos and slogans and even pictures of memes that will promote coffee. Thus, there is a chance to realize the potential of this product even in the most skeptical market.

Well, we have already talked about the problems, methods of solving them in the past and in the present. Now it's time for a flight of fantasy. The author's analysis is exactly the stage when we include all our creativity, use creative thinking and think. We are thinking about how this concept can develop. How will different social groups perceive it, how will women react, and how will men react? How will people from different regions perceive it? What and how can affect their perception? What pain points might the project have in the future? How can society perceive this? You, as the author of this brief, voice your thoughts, but on the basis of the problems that you have revealed before. At this stage of the author's analysis, you give free rein to your imagination and creativity. Here it is important to let go of the critical and start creating. And it is not at all scary if there is something funny, unrelated or maximally illogical, delusional. It should be so. Then from all this truth will be born. Then everything will be eliminated by brainstorming and only that with which it will be possible to work productively further will remain.

Analyze the concept, chosen for training.

_____

_____

_____

_____

_____

_____

_____

_____

_____

We analyze the brief using exercises you are already familiar with, which is the study of paradigms.

At this point in the brief, we need to analyze the different paradigms of the concept: creators, inventors, innovators, consumers from different times, etc.

Think from different angles! What if the former is right and the latter are wrong? Then consider vice versa. Then we acquit all parties, and then we blame them. Just by the exercise that we have already done and know how it works, we shift it to a specific concept.

Exploring paradigms

(The paradigm of the powers of the medieval state)

Here we develop the story that informs us that many countries' governments banned coffee along with drugs. Coffee was traded by pirates and smugglers. But at the same time, people still drank coffee, and the coffee trade developed, even when it was illegal. The government lost the ability to tax this product, and instead spent a lot of money to keep this product from being distributed.

In the end, the use of this drink did not lead to the same consequences as other prohibited substances. In trade, coffee was used as currency, other goods were exchanged for it, and this was also prohibited by the authorities, since they could not influence the economic side of the process. It was easier to allow the sale of coffee and collect tax on it.

Explore the paradigms of your concept.

_____

_____

_____

_____

_____

_____

_____

Neural boom and the connection of everything

We take coffee and the concepts closest to it, we begin to confidently link them together so that we get interesting related concepts: a can, a cup, a store, a coffee shop.

coffee and shop = coffee fair or shop on wheels, or a coffee-car, or a kiosk in a mall that sells coffee.

coffee shop and can = coffee shop or coffee machine shaped like a coffee can.

jar and cup = jar you can use like a cup or a coffee cup. There may also be a ready-made cup of coffee, portioned, into

Neural boom and the connection of everything

which you just need to pour water. Probably all the same plastic or paper.

There can be many more of them if you combine them with each other. Next, go to the next step and write the story, which involves all of the initial word - association: "Coffee lived in the store and always wanted to get to the coffee shop. The coffee was resting in its cozy can and imagining how couples in love would rejoice in its aroma, drinking it from beautiful cups." There can be many such stories, and these are ready-made ideas for advertising campaigns or other company goals.

Turning to the story exercise. We take a concept, link it with other concepts that just come to mind. They can be close or distant. If we analyze several concepts of one project, we link them. But we focus on the first of them. For example, let's analyze the concept of mineral water.

There will also be a concept of a bottle, deposits and caps nearby. But we are focusing on water, so everyone else will be subsidiary. We don't have to have the same story for all concepts, the stories should be different - and each with an emphasis on one, specific concept.

As a result, we will have a lot of ideas that we can then voice and demonstrate to our client.

Go through the exercise using your concept as an example:

Describe the closest concepts to yours:

_____

_____

_____

_____

_____

Combine separate concepts into related ones:

_____

_____

_____

_____

_____

_____

_____

> **Go through the exercise using your concept as an example:**

Write a story based on related or separate concepts:

Summary

For the next point in our brief, we need to see the big picture, with all the data collected and our early recommendations, but since this point has a lot to do with innovation, it would be right to break it down here. It's important to say only one thing! When you know solid information about anything, when you've spent enough time generating ideas for that concept, you need to take a break and take stock. And in this case, imagine how this concept will function in the future, gradually developing over the next few years. We fill out this point of the brief at the very end of the concept book, but I cannot attribute it to the conceptual innovations that we will study further, so I publish it here.

Concept development paths

This point is the very last one. I will hint what should be written here. After all, it is the final one, although it is included in the base of the main ones.

In this one, we need to include not only imagination, but also a skeptical attitude. To predict how our concept will develop, you need to rest, re-read everything written and draw conclusions. Offer the client what he can do now so that this concept in his hands will be relevant in 5 or even 10 years.

Often you will come across such a story when the concept has already outlived its own, but those who work with it will think that it is not so. They will consider difficulties temporary and not even suspect that soon their product will not be needed. In this case, our task is to propose a modernization of the concept so that the new proposal will itself push the old concept off the stage. Find the innovation that will help him to be reborn and be relevant for more than a dozen years. Although probably this will be in a completely different form or a different concept.

Reflect on the developmental paths of your concept.

What will remain unchanged in the concept?

_____

_____

_____

Reflect on the developmental paths of your concept.

What will disappear in the concept within 5 years?

_____

_____

_____

How will the project fix its shortcomings?

_____

_____

_____

How does the project use its advantages?

_____

_____

_____

> **Reflect on the developmental paths of your concept.**
>
> In what form would the project be outside the world as it is understood?
>
> _____
>
> _____
>
> _____
>
> _____

§4.2 Visualization is everything: document design

Here we will talk about how best to arrange the document that you will receive after working out a point brief, and in what formats it can be implemented for your client. There will be a lot of them, they will be interesting, but rather blurry.

After all, if I tell you that there is the only correct way to convey the final information or result (for example, a presentation), then this in itself will be a limitation. And there can be no such thing in the world of conceptology. Because conceptology is not aimed at restrictions, but, on the contrary, at development and going beyond the framework.

I can give you some useful tips that your clients will love.

1. Consider what your client needs. What media does he use most often? Reading or watching a video?

2. Observe the client: is he an auditory or a visual learner? How will it be easier and more comfortable for him to perceive information? Ask if he reads books, listens to them, or prefers to watch documentaries based on these books.

3. Appreciate your client's time. Anything that can be simplified, simplified, and what can be reduced. You shouldn't make huge amounts of text where you can get by with a couple of blocks of infographics.

4. Include imagination and combine. Nobody limits you; you can choose several media and create something of your own. Every company always has a lot of documents, so those for whom you will work will be familiar with this. But so that your presentation does not turn into another folder of papers, it is worth being smart.

5. Don't use too many terms and scientific words. Everything that can be translated into simple human language should be translated so that it is clear to both the professor and the grandmother at the entrance.

Companies keep all the most important matters recorded on paper: charters, internal regulations, office work documents, contracts, instructions, commercial proposals. But reading these documents is a torment. As a rule, such documents are created according to the same scheme. Plan. Introduction. Basic provisions. Conclusions, results. Bibliography. There are rare exceptions when documents are drawn up in the form of banal presentations (for example, a set of corporate culture expectations or some situational local documents, often related to the activities of staff), but they also do not look interesting and impressive. There is no escape from the bureaucracy, but we can always change what is in our power. Why would a company add another document folder if it doesn't need it?

You can present a concept book in completely different ways. For example, make a video presentation, where everything will be decorated with documentary footage, valuable images, beautiful slides and visual pictures. Or create a whole movie with the information you need. You can create a special application or an interactive game (if there is a resource and budget), which can even show several results, depending on what action the client chooses (for example, as in the game The Witcher or Cyberpunk 2077, where there are several different endings. You can make a glossy magazine with beautiful and colorful concept photos (thus, a stack of interesting magazines about the company's concepts will appear on the tables of the company).

You can create an electronic concept book with embedded gifs as well as clickable links that will link multiple other concept books together and be complementary. You can shoot a short clip explaining the main points from a regular balance presentation. But in reality, there are no restrictions. The best is to do what is convenient for you and your client.

§5 CONCEPTUAL INNOVATIONS

When working with innovations, it is essential to understand that some of them would still have happened by themselves. Historically, at a certain point in time there was a need for innovation. This is not something unusual, but something most often what everyone was talking about, but somehow their hands did not reach to realize. The innovation had no other option for how it could happen. For example, before work, people went to drink coffee in a coffee shop, at some point they wanted to take coffee with them and drink it on the way to the office.

People came up with paper cups, but they've already been used, just for other purposes. When people did not have money for expensive dishes after the war, the business of plastic and paper cups skyrocketed. They were cheap and could be used many times. People from these times saw this as a good alternative.

But then they did not think about comfort, but rather about covering basic needs. However, coffee lovers picked up this idea and used it in their own way - to enable people to drink coffee on the way, without interrupting important matters. This is just an example of how an innovation that already existed works. Everyone has been talking about it, but someone brought it to life and finally started doing it.

Conceptual innovation requires an effort

Conceptual innovations (the individual must spend time and effort to work with this methodology) are needed in order to make the concept closer to fruition with the consumer or distribution channel. They take the negative perceptions and attitudes they start out with to become actually positive by the end. They convey the positives that the concept brings to the audience and turn the minuses into pluses. Here our task is to turn vulnerable and negative points into advantages. We need to take the most painful points, because of which the consumer does not want to try a stand-alone product or its basic concept and turn them into strengths. Once again, the vulnerability is what causes the majority in the market not to buy a product. These vulnerable points need to be turned into positive decisions.

For example, few people want to buy decaffeinated coffee, because they believe it's all chemistry, and real coffee cannot be decaffeinated. No one provides evidence that this is so or not. This is a vulnerable point for those who sell decaf coffee. They have their own target audience who would be happy to buy decaffeinated coffee (for example, because they cannot be caffeinated for health reasons), but they are convinced that there is such a terrible and deadly chemistry that it is better to suffer from caffeine or even use chicory even though there may not be any scientific evidence of this. This is where we need to work through this vulnerability by conceptual innovation.

Another example is eyeglasses. The vulnerable point is that glasses are often prescribed not only for poor eyesight, but also when the vision has slightly decreased. It is also necessary to wear glasses when the eye muscles are especially tense: for example, while driving or while reading. Then people begin to tell you, that if you have already started wearing glasses, your vision will inexorably fall, and you will almost go blind due to the fact that you use glasses, not being almost blind like a mole. There is another vulnerability in the concept of wearing glasses and the concept of the glasses themselves: if a child needs to wear glasses, they can be teased and bullied at school. Such examples need to be looked for on the largest scale, and after a successful hunt, begin to turn these triggers of bias into the desired result.

Example of large corporations

The modern world is rapidly digitalizing, making us obligated to stand out. Thanks to the help of thought leaders, we have various projects that grab the users' attention, making their lives more interesting or comfortable. Let's take an example of the history of the well-known social networks Twitter, Instagram and Snapchat. They took turns taking each other's audience in a variety of ways. Basically, these methods were related specifically to the functionality and concepts used, in the minds of users, especially if we go a little deeper into the tasks that the above-named social networks perform.

• The idea behind the Twitter app was to enable people to share their thoughts out loud with each other using short messages. The original concept appears to have been borrowed from pagers. A person could react to some well-known news, wish good morning to his friends, or report important news that everyone around him will certainly know about. News channels could quickly spread the most important news by adding a link to it or a photo from the scene. The main thing is that Twitter met the need for people to voice their thoughts and, of course, to joke for all sorts of reasons. However, this platform still has its own weight, even after the entrance of other competitors. Many politicians, stars, official resources, and media people still use it to convey their news or thoughts to the world, creating a real resonance with their respective audiences.

• But when Instagram appeared, and the first reaction was: "Why do we need an application where we need to display photos and sign them, if there is a twitter, where you can write text and attach photos?" But the difference was huge. After all, this project was originally about the fact that you can show your life with the best pictures to your friends. Moreover, you can see their instant reaction in the form of a like and a comment. And you don't have to sign your photos, because the concept of this application solves the problem of exchanging emotions, not information. The difference is obvious, isn't it? "Stop a moment, you are wonderful. I'll take a picture of you and show you to all my friends! " - this is how Instagram could be characterized at the beginning of its journey. YouTube took a different approach in that it was simply a repository of data that all friends and acquaintances with the link could access. In the same way, this website has become a commercial platform for bloggers and influencers. But now it is the most powerful unit for marketing and sales, a platform that engages millions of people. It is a place where everyone tries to expose or fabricate their own interesting life, and just keep an eye on others.

• Snapchat broke all templates. It seemed that Instagram was going through its last days and would soon become as uninteresting as Twitter turned out to be for many after the appearance of Instagram. Snapchat burst into the market with a very loud claim to the functionality of the application. First, there were masks that were applied to the face. At that time no one else had ever suggested anything like it. Second, they proposed a video format that was stored for only 24 hours. As a user of both of those at that time, I can confidently say that Instagram was seriously losing out to Snapchat, because it didn't have such functionality. At the time, Instagram only provided the ability to post photos and 15-second videos to the feed. Nobody even thought about masks, or stories, and even more so about IGTV. But what happened? At some point, Instagram updates were released, with new effects, and a story for 24 hours, and even the possibility for videos up to a minute in length to be displayed in the feed.

It is no coincidence that I told this short story, because exactly what happened in all these cases is the conceptual innovation that we discussed. As a result, we will spend more time on Instagram since this social network responded in time to the threat to the conceptual component of its brand, and also selected the best attributes and functions from its competitors. But it did it elegantly, on a large scale and without copyright infringement.

From the very beginning, Instagram has been a trendsetter. A trendsetter is one who sets a trend, establishes new fashion and trends in a new direction. It connected both what people wanted from Twitter and what Twitter lacked. That is, Instagram retained the concept of a short news story, but at the same time incorporated it into a completely different concept of a photo gallery with access to all friends. However, according to my observations, at first no one took this new format of the social network seriously. After all, there was Facebook, and this aggregator was then a powerful platform for social interaction between people: communication, exchange of photos, communication in groups. Moreover, it was a role model for other foreign copies of this social network. Thanks to Facebook, as soon as there was any need in the community, it was immediately picked up and implemented into reality.

But Instagram initially offered itself not as a social network, but as a convenient photo gallery in your pocket, which you allow your friends to see, and even get a reaction from them.

When Snapchat began actively attacking Instagram regarding its functionality, it solved this problem by simply providing millions of current users with the same functionality, and it did it really fast. In addition, in the Instagram format, this functionality has already become a story much more interesting than that of Snapchat. Instagram used masks not just for entertainment, but as an element of sales, brand awareness and promotion. Stars were the first to test it- they invented and launched their masks.

For example, Rihanna has a diamond mask, and Kylie Jenner has different shades of her lipstick. The idea was picked up by shops and created virtual showrooms. Just think, now, thanks to the masks, you could literally try on something: glasses, an item of clothing, a hat.

And try to tell me that this is the job of marketers! No, this is the work of concept geniuses - product managers, developers, lawyers and analysts. Marketers, of course, but with a very unconventional approach to marketing. This is how Instagram competently picked up and developed this trend. And then — more. It kept improving and improving its mechanisms. It increased the amount of time for stories, added the ability for everyone to create their own masks (previously, this function was available only to selected testers or those who submitted an application for about a year), then Instagram created the ability to conduct and stream live broadcasts, created IGTV to save the best video content and upload useful videos.

I think I just confessed my love for Instagram. However, this company has never broken from its original concept - a gallery with text additions, which can be viewed by friends and are shared for their reaction.

§5.1 By reinforcing the negativity

In 1993, in the USA, the city of Seattle, there was news on TV that a married couple had found a syringe in a Pepsi can. A few days later, another complaint of the same content was received. Over the next 24 hours, a wave of news coverage swept across the national media about syringes in cans of the world-famous beverage.

On the same day, a local Seattle bottling facility for Pepsi-Cola launched its own investigation to investigate the cause of the problem and then respond to the media and the public.

The Seattle-based Pepsi-Cola company has allowed local media crews to enter production so they can show the latest high-speed canned beverage line. Press releases were also issued, in which the consumer was assured that the company would find the cause of the accidents. The PR team has developed news videos, press releases, consumer arguments, photos, and interviews. All this had to reach the respective audiences and quickly stop the panic.

The anti-crisis team recruited Robert Chang Production, a longtime manufacturer of television products for Pepsi-Cola. This company had to create a video material that flawlessly affects the audience, illustrating the message from Pepsi-Cola. In this video, the "syringe hoax" was fully developed. Arguments - such a number of claims, presented in so many different places at the same time, cannot be based on real grounds purely logically.

The Pepsi-Cola CEO spoke on news programs on all major channels to announce that the company was "99.99% confident" that this could not happen at Pepsi-Cola facilities. Although this crisis cost the company a decrease in sales of $ 25 million, by the middle of summer Pepsi-Cola made up for this loss. The summer season ended with a record level of sales over the past five years, 7% more compared to the previous summer season.

Algorithm for enhancing negativity

A new student named John comes to school, and rumors started to be heard that he was kicked out of his previous school for fighting. There is no accurate information about it, but other students are rather suspicious nevertheless. (there is a parallel here with black skin color, harmful sugar, mineral water, and so on).

Then one day, one of the school employees comes into each class and reports that John broke the nose of the local star baseball player Peter. Peter was taken to the hospital by ambulance, and John was taken home by his parents. Everyone starts to hate John. In the corridors one can only hear: "Well, he just came to our school and is already creating trouble. He does not belong here."

But by the evening the school principal comes to each class. She apologizes that the wrong information was reported and now tells the real story. In the morning, John was going to school and saw Peter breaking the phone of the most popular girl in the school - Lisa. Peter was dragging her on the grass by her hair and shouting at her. After asking him to stop this, John defended her and ended up breaking Peter's nose in the process.

And suddenly, everyone started to adore John. He became a local star and an icon of justice in this school. His heroic actions will remain first and foremost in everyone's minds. If anyone doubts, John is now known for his heroic act, with the shame of those who condemned him without knowing the real truth. All of Lisa's fans and Peter's haters will stand up for him.

An innovation on the example of turtle soup.

Usually, a turtle soup dish is not commonly prepared in every home. It is rather considered exotic and maybe even extravagant. But if we say and prove scientifically that turtle soup cures some serious illness, for example, cancer, or if such a soup is consumed once a month, then the probability of contracting cancer is reduced to zero, the world will immediately turn upside down.

Everyone without exception will want to eat turtles, they will start raising turtle farms, new businesses will immediately appear, and turtles will become more popular than cows or chickens. In the meantime, turtle soup is of little interest to anyone, since it does not represent a value that would interest the world. Moreover, for many, since this dish is disgusting, the question must be asked: what would be needed for this kind of innovation to take place?

By reinforcing the negativity

First, we hear that decaffeinated coffee is nothing but chemicals and should not be drunk. People begin to believe this. We are waiting for the media and people in their kitchens to start discussing this, along with opinion leaders who start communicating to their audiences about this, and their comments are created from controversy.

But then someone does research and it turns out that decaffeinated coffee is still the same coffee, only produced in a slightly different way. Soon, this information is spread over the Internet, and everyone is carefully studying it. It turns the whole picture upside down.

Now, even those who were avid fans of caffeinated coffee are starting to wonder about their coffee. They wonder if

By reinforcing the negativity

maybe they shouldn't drink caffeinated coffee, as it is harmful for the heart and blood vessels. It's proven now that caffeine-free isn't so bad. And now all decaffeinated coffee sellers are jubilant and counting their money.

Analyze your concept by reinforcing the negativity.

What flaws or myths can you reinforce?

_____

_____

_____

_____

How can you knowingly debunk them or prove otherwise?

_____

_____

_____

_____

> **Analyze your concept by reinforcing the negativity.**
>
> How can you use it in business?
>
> _____
>
> _____
>
> _____
>
> _____
>
> _____
>
> _____

§5.2 By reverse bullying

The world has been abuzz with marketing wars for the past few decades. This is a story about when two giants from the same sphere are at war with each other, conduct daring advertising campaigns, boldly ridicule the opponent's shortcomings, try to quickly introduce some technological innovations, and want to be more creative. But here it is important to remember that marketing wars are about business, but not about innovation, because they create a community of those who support one and reject the other. Although they concern not only companies, but also adherents of diametrically opposite things. For example, meat eaters and vegetarians, or lovers of alcohol, tobacco, and healthy people.

Three whales, on which rests concept vulnerability: stereotypes, prejudices and bullying

There are many examples of those who prefer to use Apple technology, and those who consider themselves the ambassador of Samsung; those who love KFC more than McDonalds, or vice versa. But we are interested in the essence of conceptual innovation, which lies in the fact that in the process of disputes between the two camps, what makes these giants move forward is born. Something that motivates companies to fight for the hearts of their users and constantly come up with how to improve their product, how to interest and manipulate those who are the target audience.

Traditionally, let's take a look at the example of vegetarians and meat eaters:

Meat eater's arguments
You eat only plants; you do not get vitamins. You cannot do it this way.
Not only do you eat one type of plant yourself, but you also impose your way of life on the child. And he needs protein.
Fields and plantations suffer because of you. Even animals suffer because you take away their food.

Vegetarian's arguments

There is nothing healthy about meat, now it is synthetic and stuffed with antibiotics.

It is the killing of innocent animals.

The fact that you eat meat is not your choice, but the ideology of huge corporations imposed on society, which cannot survive otherwise. You are their slave.

It is through these discussions that innovation is born. Vegetable meat appears, which should satisfy the needs of both the first and the second. The interface in iOS and Android is becoming more and more similar. Burgers appear at KFC, and crispy wings at McDonalds. Everything in order to entice the audience of a competitor.

In order to keep their own, models are created with a waterproof and powerful battery, a camera and software adapted for social networks are being thought out, etc.

By rejecting bullying

The main competitor to coffee drinking is tea drinking. The main competitors use the fact that they popularize tea ceremonies, they like to have a beautiful time with tea. They argue that tea is healthier, and rightly note that drinking tea makes it easier and more pleasant to sleep at night, and in the morning, makes you feel a surge of energy.

We would like to point out that this innovation should focus on decaffeinated coffee, which can be drunk even at night. This is a strong argument for the late-night coffee argument. In addition, we can develop caffeine-free herbal coffee and even a soothing coffee before bed. Natural tea pigments are more likely to stick to tooth enamel than coffee pigments.

For daily use, this type of innovation doesn't work. Researchers from Surrey County University in the United Kingdom have confirmed that both drinks are equally beneficial for focus and concentration throughout the day.

Add your idea about coffee drinking:

Analyze your concept by rejecting bullying

Seeing the concept variation from the first side:

_____

_____

_____

_____

Seeing a concept variation from the second side:

_____

_____

_____

_____

How can we combine their arguments in a new proposal?

_____

_____

_____

_____

§5.3 By the miracle effect

Right from the start, this method of conceptual innovation sounds intriguing. This is because it works with that part of the human conscious and unconscious that always responds to magic, beautiful stories, incredible adventures and absolutely everything that is connected with the feeling of the presence of miracles.

This innovation communicates directly with our inner child, who has not yet "spoiled" his attitude to the world with various experienced fears, pains and losses. Only a child can believe in a miracle without any real evidence, being 100% sure that it is true. Take a moment back to your childhood or adolescence: remember how you hoped that one day an owl would bring you a letter from the school of wizards; or how you tried to stay awake until New Year's Eve morning in order to see Santa Claus with your own eyes, wrote letters to Saint Nicholas about the desired gifts, or left a carrot for a flying deer and milk and cookies for a magical grandfather during the night. So now you will learn how to improve virtually any concept, using these feelings for the benefit of the concept itself and all its users.

Fantastic real life stories

Even though we are all getting older, somewhere in the depths of our souls, belief in a miracle lives in everyone. The essence of innovation through the effect of a miracle is precisely to hook these strings of our eternal inner child. But the great thing about this innovation is that it is not fiction. In the end result, the miracle actually happens. You are the wizard in this process! You can often find inventions that previously existed only in futuristic books and science fiction films. We are already using them in real life. For example, in the books of science fiction writers of the past century, video communication devices were often used. These were not just common everyday phones, but during a call you could see the other person in real time! Back in the 1990s, it seemed like this thing would be impossible, but everything has changed now!

We can make calls using a piece of metal in our hands and see each other through a tiny black spot on this device! This is the miracle that passed into everyday life and has just become a part of it. Another example - in books they often wrote about a new format of television: that is, you do not watch what you are shown at the exact time that they schedule, but you yourself choose what to watch and when to watch it.

If you told someone at the end of the 20th century that they would walk around with a thin metal tablet on which you can watch thousands of movies and TV shows of your liking, they would tell you that you were greatly impressed by Star Trek. But in 2021, there are dozens of tablet varieties that have millions of hours of free content on YouTube, Netflix, Amazon and others.

The main secret in creating a miracle

The older a person is, the more negative experiences he goes through. We cannot change this fact in any way. However, the more a person has experienced negative experiences, the more shocked he is when he cannot explain an elementary card trick. What can we say about when real magic happens before his eyes? **"The adult no longer believes that a miracle can happen, but the inner child continues to hope that it is possible." ©**

It is in the creation of inexplicable magical solutions that the main role of innovation occurs through the miracle effect. The moment when people are disappointed and have lost their faith in a miracle is the most ideal time for any concept to show them this miracle and give it for external use.

It was on this that Steve Jobs once played, and then he continued doing so all his life.

He made more than just a computer. Computers used to be completely different: they looked like a keyboard with an operating system. Steve also produced these at the very beginning of his journey. They had to be connected to a TV, and this was the only way to work. Such a computer helped to calculate math problems, learn geometry, create databases, and even play primitive games very quickly. In the days of Jobs' contemporaries, it all looked like an endless stream of text divided into tablets, but Jobs was one of the first to create a graphical interface using a computer mouse, a set of text fonts, and a drawing program.

People knew it was cool when you had a personal calculus robot at home that instantly copes with even the most difficult tasks. It's really great to have on your desktop what used to take up entire rooms in factories. But no one could even dream that the computer would turn into a part of your consciousness, which would beautifully arrange files into folders, control the cursor, and even more so draw in different colors on non-existent paper. Sounds like a fairy tale, doesn't it?

At that time, graphic designers were already using a mouse. They guided it across the screen and when a button was pressed, it would have just started drawing. However, it was not a cursor for control, but rather a tool that transfers lines and points of different shapes to the monitor screen.

Jobs saw this one day and said, "The future of computers is the mouse. You will see, time will pass, and people will not be able to imagine a computer that is not controlled by a cursor." He said that the mouse will not be used to exclusively draw straight lines, but in order to generally use the computer, and so it happened. Try to imagine your computer without a mouse or touchpad. This is how Jobs pioneered innovation through the miracle effect. In fact, it was a very simple innovation compared to his next innovations using the same effect.

Steve, of course, has always been a master of presentation. Therefore, people often experienced emotional shocks at his performances. He knew what levers to push and when to make a presentation so that the effect would evoke the most vivid emotions!

He realized that you could play well to what people want. He investigated trends, listened to human desires - and was always one step ahead. When marketers came to Jobs with a statement to conduct a survey of the population, he replied: "This is absolutely unnecessary, because people themselves do not yet know what they want." People perceived Steve as a genius with a vision of the future. But in fact, he just found the miracle that people are waiting for and did everything possible to bring it to life. Brilliant, isn't it?

In 1995 Steve said in an interview: "You know, it hurts Apple badly that after I left John Scully had a very serious illness. I've seen it in others too. When you think a good idea is 90% success. As soon as you tell everyone what and how to do, of course, they will go and do it. The problem is that it takes a lot of effort to turn a good idea into a good product. When you improve this idea, it changes and grows. It never comes true as intended. Because you learn more by going into the details and understanding the big trade-offs you have to make. It's just that some things are not subject to electronics. They are not subject to plastic, glass, factories or robots. Once you understand what it means to create a product, you have to keep in mind five thousand things, understand these concepts, combine them together and continue to observe them in new forms in order to achieve what you want. Every day you discover something new, be it a problem or a chance, to combine these things a little differently. And this process is real magic!"

But what else, if not the effect of a miracle?

In my opinion, now this is no longer there. Apple now gives people what they expect: three cameras, a thin laptop, a waterproof case, even better design, even thinner and lighter options. The reason for this was the absence of the company's chief conceptologist. This has happened before, when Steve was first removed from his own company.

Having created a similar company NeXT, which Apple subsequently bought out, returning to Steve the position of president of the company, he gave the above interview.

In addition, he commented on the state of Apple at that time, which is very similar to its current state: "John Scully came from PepsiCo. From a company that hasn't changed its product for at least ten years. The most they can change is to suggest a new bottle size. But if you're a salesperson, you can't really change the course of a company.

So, you will only be surrounded by salespeople and marketers. They think about how to sell products to increase the value of the company. This is probably enough for PepsiCo. But if that happens at a tech company, they become a monopoly. But what do you do next? Better copy or better computer? "

"The company forgets what it means to make good products. The geniuses who work on the products are left out by the people who run the company with no concept of distinguishing a good product from a bad product. They usually don't know how to really help users and are just trying to find the best way to become the most famous company in the computer industry."

Now the way of this company evokes the same feelings when they shoot a film based on a famous book. Most viewers have already read the story and imagined what it might look like. They go to cinemas, but there is no miracle. After all, everyone imagined everything differently and often even better than it turned out in the film.

Obviously, the maximum success is to get on the same wavelength with the reader and shoot everything the way he imagined it. Or hope that after all, most viewers have never read the book. But in the case of Apple, this is virtually impossible, because people have been using their products for a long time and assume, perhaps, in the next versions of their inventions.

People are constantly a little unhappy. Because in their heads everything is different, and the only way to surprise them is to provide what they need, what they could only dream of. Along with the departure of Jobs, innovation through the miracle effect also left the Apple world. Moreover, now competitors are picking up trends faster. And if in the 2010s iOS and Android were different, now even their interface resembles each other.

Let's figure out how to create this effect

Just as with any innovation, we have a structure that helps us make a miracle. First, we need to find the functionality of the concept, which we encounter on a daily basis, and it brings us discomfort. What annoys you about this concept?

If people are accustomed to what annoys them, they no longer hope for a miracle. Our task is to find as many of these "bugs" as possible and use them to improve the quality of life of the person using the concept.

By the effect of a miracle

What's annoying?

- You cannot drink coffee before bed because you will definitely not fall asleep for a long time.

- It is impossible to understand how much caffeine they need in their bodies right now. Is it even necessary to drink coffee at the moment?

How can we fix it?

- Make a special herbal coffee that tastes the same but doesn't affect sleep in any way. It even helps you sleep.

- Create a device that will determine at a specific moment how much caffeine you need to consume. Perhaps it will look like a bracelet.

- It is necessary to release an original coffee machine and sticks filled with coffee with different caffeine content. When the machine receives information about the required amount of caffeine, it makes coffee that will only improve the user's condition and definitely will not harm him.

Analyze your concept by the miracle effect

What's annoying?

_____

_____

Analyze your concept by the miracle effect

How can we fix it?

§5.4 By deliberate reputation enhancement

Reputation and image are everything, especially in the modern world, when you can find information and answers to almost any question. The most negative news spreads with lightning speed. The Internet is full of customer reviews and reactions to the slightest negative moments.

For example, if a seller of a company treated you in a store disrespectfully, and you wrote a review about it online, then there will certainly be proceedings within the company, as well as an analysis of the situation why the seller did this. Even if this happened in a huge corporation, such tendencies will be recognized, and the maximum number of specialists will be involved to eliminate such a problem. And then, perhaps, a revision of work standards, reprimands will result and, most likely, you will also be given some kind of present or a discount. This is the practice of many brands that value their reputation.

Because now, in order to be loved, respected and wanted by your corporate or personal brand, you need to provide not only a high-quality product, but also a first-class service. It doesn't matter if you are a beer maker, a local bakery, a writer, a singer, or an actress. People pay to use your product, look at you, or listen to your voice.

How does this improvement work?

Innovation through knowingly improving reputation is when they do everything to improve their reputation before something irreparably bad happens. Everything seems to be logical. But if the reputation is already very good, then why improve it? The easiest way to explain this innovation is with the example of Emilia Clarke (Daenerys Targaryen in Game of Thrones). Long before the release of the last season, she already knew that she would be perceived negatively, because in the last episode, the character played by the actress turned out to be extremely negative. In the world of movies, it happens that the audience identifies the character directly with the personality of the actor. It can even ruin the entire career or strongly tie the actor to the role. Take actor Tom Felton, who played Draco Malfoy in Harry Potter, for example. For a very long time he was perceived exclusively as a negative hero, an actual villain. In an interview, the actor said that children, when seeing him in the city, were afraid to approach him.

Emilia was wise to use innovation through knowingly improving her reputation. About a year before the release of the last season, she opened her own channel on YouTube, began to meet people and show herself as she is - a sweet and pleasant lady. Then she began to actively record videotapes. Including, along with other actors of the film, she appears in funny videos or, disguised as other characters in the film, walks along the main streets of New York. Emilia acts in sensual and romantic films, plays good positive characters, jokes a lot and participates in various talk shows. The audience perceives this simply as the usual actions of a public person. At the same time, the audience admires her. But a year later, as planned, everyone is shocked by the latest episode of "Game of Thrones".

She innovated her own personal brand, and improved her reputation even more in advance, knowing that soon something will happen that will definitely worsen it. There are many ways to do it. The same with brands: the more they make themselves friendly, sympathetic, kind, charitable, funny, in contact with the audience and close to the people (not in words, but in actions), the less they are afraid of the blow to their image or reputation, which will definitely happen.

In this case, people themselves will justify some kind of brand mistake. And to the terrible news, which will definitely worsen their reputation, they will say: "They are so good, this is an accident!" or "They do so much for people, everyone can be wrong. There is nothing wrong with that. "

What is worth doing to repair a damaged reputation?

• Conduct charity festivals
• Share gifts
• Help pay for treatment
•Create funds to help those in need
• Fulfill someone's dreams with your resources.
• Arrange various contests and challenges
• Communicate with your audience in person.
• Do everything to declare yourself as loudly as possible.
And spread the good word!

What is innovative?

The point is that we still work with concepts, not with a brand. This innovation did not appear here by chance. After all, while this is a great idea for marketers, and they are already in a hurry to bring it to life in their companies, conceptologists can use this knowledge in a completely different way.

When working with a concept, we need to focus on how to improve the reputation not of the brand, but the concept itself. After all, the concept also has it, and it may well be both good and bad. Have you ever wondered what the reputation of the concept of a funeral in black and a funeral in white has? Have you thought about the reputation of the executioner profession in the Middle Ages? And what about the profession of a judge? And who was worse? This innovation gives us the opportunity for the first time to ask questions about the reputation of a concept with different strata of society, and then to predict where the reputation will definitely deteriorate and to carry out exactly the same manipulations with the concept that any experienced marketer with a brand will do.

By deliberate improvement in reputation

We know that coffee can help in some situations. For example, not falling asleep when you have to work hard, or waking up and feeling cheerful when you haven't slept for a long time. We use exactly these advantages, take them into circulation and, on their basis, arrange high-profile events.

- We bring coffee from different countries to expand and develop the culture of coffee drinking. Be the first to introduce the consumer to the experience of coffee drinking in other countries. We provide special coffee prices for university and schoolteachers and ID students.

- We bring coffee to nursing homes and hospitals to make it easier for doctors to work night shifts, and patients with low blood pressure can use our coffee.

- We show coffee in films and make it an attribute of goodness: for example, when a detective needs a boost to figure out the riddle, and he explores it all night, drinking coffee. And then it dawns on him, and he finds a solution.

By deliberate improvement in reputation

Now we adapt this even more to the state of the concept. And instead of the brand of coffee itself, we are promoting coffee drinking. Not just our brand, (marketers will do it), but the coffee ceremony. Perhaps the same detective in the film will devote 15 minutes every morning to the daily ceremony with his chips. And the teachers will ennoble coffee drinking at meetings or on breaks and will not just drink coffee but will treat it with a special approach. Students will not do this in parallel with writing essays but will allocate time for this ritual. This means that we sell not only coffee and not only many accessories for coffee drinking, but also the very ideology of drinking coffee not "by the way", but deliberately and enjoying the process as much as possible.

For example, in one of the branches, they found that rats walk on coffee and leave traces of their physiological activity there. We were the first to know about this, and our local representatives have already done everything to fix the problem. But one of the former employees of the branch has already told the press. The press will soon tell everyone about it. Instead of worrying, we can simply assume that we have improved our reputation sufficiently in the past and in response to the news, we will write that this is an isolated case. Of course, users will think: "Yes, everyone has it, they overlooked it. But this is only one branch, not the whole company. Moreover, they have already closed it and fixed everything. And their coffee is delicious, and inspections are constantly conducted. The president of the company said this morning that they checked all the other offices - and everything is fine there. So, not just in the factory of the rat, but in the entire company. So, there are definitely no such incidents when packing the coffee. " This is exactly how innovation works through a deliberate improvement in reputation

Analyze your concept by improving your reputation

What are the benefits you can use in advance?

_____

_____

_____

_____

_____

How can these benefits be realized?

_____

_____

_____

_____

_____

_____

§5.5 By collaboration

The lexical meaning of the word collaboration (from Latin con - "s", laborare) - cooperation, interaction, cooperation, joint activity. In simple words, this is a joint activity of two or more parties to obtain certain results. The parties involved in a collaboration are called collaborators. We can define several types of concept collaboration:

When you collaborate on two disjointed concepts for the purpose of ideological addition.

For example, take sneakers and Charlie Chaplin, who in the movies wears terribly uncomfortable, huge and funny shoes. And then we say: "Look, Charlie Chaplin wore such boots, and he was terribly uncomfortable. But if he walked in our sneakers, he would not feel an ounce of discomfort. " And you release shoes of this format, which borrowed some of Chaplin's features.

We are used to this innovation, it is often used by marketers, but although it uses different concepts, it can hardly be called conceptual. It just is, it allows you to look at familiar things from a different angle. But it doesn't carry any qualitative changes in these two concepts. This type of innovation just needs to be known and understood. In the concept, in fact, nothing has changed much. What has gotten better? The appearance has changed.

There are many examples of such collaborations in the world. In the fashion industry, Disney and Dolce & Gabbana are the most sought-after brands to merge, or Dolce & Gabbana and home appliance manufacturer Smeg; the same "mouse castle" and Alexander Terekhov; the case when after the release of the film "Maleficent" a collaboration product appeared in the form of a whole line of cosmetics from MAC in the style of this film and the main character. For the same event, you can remember the new series The Simpsons MAC.

When collaborating with others, we improve the concept

Back to the Future featured a Nike sneaker with self-tie laces. A dozen years later, this company released the same sneakers, when such a fantasy technically became a reality. They did create a shoe that was an improved version of the previous shoe (the laces were really tied), and at the same time it was a collaboration with the film.

When collaboration takes place within the concept itself

Lucky Strike was originally a factory that made cigarettes from different varieties of tobacco for different brands according to specific recipes and standards. But once the workers went on strike - they were not satisfied with the working conditions and wages. They put forward a condition for the management: if no one listens to them, they will mix all the varieties of tobacco. Consequently, the company will lose a lot of money. An entire consignment of tobacco will be irrevocably spoiled. As a result, the authorities ignored the threats, and the strikers still mixed the tobacco.

The strike died down, but the owners had to figure out how to fix the situation. They decided to find out what came out of the tobacco hodgepodge. This is how a new brand of cigarettes appeared, which was named Lucky Strike. This is how they released a new product that was not even supposed to be on the market before. So, they also backed it up with such a beautiful story.

The second example is the Kievsky cake. There are many stories of its origin. The most interesting is when the pastry chef turned off the oven, but forgot to get the cakes, and they turned into dry pieces of dough. To avoid punishment from the management of the Soviet plant, the pastry chef broke the cakes, smeared them with cream and said that he had made such a decision to increase the assortment. People liked the cake, and literally in a few days there were whole queues for a tasty novelty. Moreover, there was a serious shortage of this assortment in the Soviet Union.

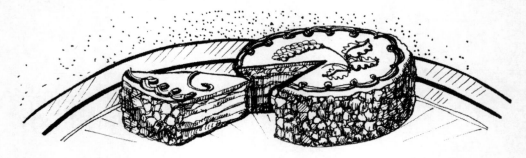

An interesting example of conceptual innovation, when such a collaboration occurs on purpose, is in the culture of hookah smoking. The innovation through collaboration made sense when different fruits were used instead of a bowl: coconut, apple, pomegranate. Add drinks to the flask - from wine to milk and original mixtures. Make special ice flasks and tobacco blends, create smoking stones for those who do not want to smoke tobacco.

By collaboration

You can also use your own concept here and collaborate with it. Using coffee as an example, we can make a collaboration with tea. Roast coffee, grind it into powder and do the same with tea leaves. Mix and get a new drink. Or mix coffee powder with cocoa.

Mix coffee with other unexpected foods that you can dry and grind into powder. As a result, not only a new assortment of goods will appear, but, possibly, a new wave of users for whom this drink will become their favorite.

Analyze your concept by collaboration

----

§5.6 By obsessive development

One of the modern laws of business says: "Stand out or die."No wonder, because every year there are more and more businesses and fewer niches. What to do in order not to burn out in the first months of its existence? How to hold your ground when you see new companies nipping at your heels? The main thing here is to be brave and not be afraid to change the concept that is familiar to you.

When it seems like there's no way out

If you or your client has this feeling, then you need to turn to innovation through compulsive development. This innovation is most often worth using when there are already many competitors in the same niche on the market. This innovation is the most offensive, because it means that your business is not in the blue ocean of opportunities, and you have to fight hard. Not only with competitors, but also with your own ego. Because the business owner needs to take a very critical look at the area in which this business operates. Honestly to tell myself that my product is the same as that of others, that at this stage my business is no different, and it has nothing special or very little. And then analyze all the "incoming data" about your state of affairs and change your positioning and concept for 360 °.

Catering services are great example. When there are several markets, and all the same:

- Selling the same products.
- Use the same shelf placement.
- Provide similar customer loyalty terms.
- Dress the staff in a similar uniform.
- Built on the same principle.

Or, for example, steak house where the meat / suppliers are the same, the idea is similar, the recipes are similar, even the music is similar. Or the same coffee shop. We noticed how many coffee houses are similar to one another: small, with the same machines, all offer "to go" coffee, brew all the most typical: latte, cappuccino, espresso, Americano. They sell cookies and health bars.T hey are also a meter apart. What will a person be guided by when choosing? Most likely, they will simply choose a closer place, or where he likes the barista better.

It is same with everything. To win over your consumer, you need conceptual change and sharp ingenuity. In this regard, a striking example is the Ukrainian city of Lviv. It is known not only for its beautiful architecture, but also for thousands of conceptually special establishments. People from all over the world come here to enjoy not only the soul of the city and delicious food, but also flashy conceptual solutions. There is every establishment or network of establishments with its own unique chip. Although, it would seem, everything is about food.

· There is one interesting place here, where they will bring you a bill for 15,000 Ukrainian hryvnias (the currency of the state of Ukraine) when you ate for 1000 hryvnias, because it is customary to bargain here. If you are eloquent enough, you can pay even less, and if you are not very convincing, you can pay much more.

· There is a place where all prices are initially × 10. That is, instead of 120 for a salad, 1200 is indicated on the menu. Here you need to know the code word or complete an assignment from the waiter so that all prices are divided by 10. But if you want, you can pay 10 times more.

• There is a coffee shop in the shape of an old mine. And there you can literally dig up some coffee. You will even be given a helmet and a trolley for the entourage.

• There is a place where everything is done in the style of a monastic dungeon. Stone walls, robed waiters. Clay crockery and cutlery forged by real blacksmiths. Soup is drunk here from cups, and meat is served on a hot steaming stone.

• There is a place where all cocktails are brought in laboratory test tubes and flasks. They are of a wide variety of colors and capacities.

What needs to be considered when compulsively developing a concept?

Introduce this concept into the user's "daily diet". Until around 2015, Los Angeles never had an instant rental electric scooter. People bought their own, and they most likely were not electric. Suddenly scooters began to appear right in the middle of the road, with an offer to rent them through a special mobile application. Then it became possible to rent a bike. Literally a year later, all my friends were under the impression that the rent of bicycles and scooters has always existed. Moreover, it is difficult to remember the times when this was not the case, although in fact very little time has passed. This can now be called a blood connection with the brand. When a brand is in the blood. But we'll talk about this a little later.

These are good examples of how compulsion innovation is applied. But in order to master it professionally, you also need to understand the so-called pyramid of branding. I learned about this pyramid from my first master, Volodymyr Oseledchyk, at Kyiv National I. K. Karpenko-Kary Theatre, Cinema and Television University.

He was in the team of founders of one of the most successful Ukrainian TV-channel «1+1», and this pyramid also has formed the basis for the creation of this channel. But we will consider it from the point of view of the readiness of all the concepts that the brand uses. Therefore, now let's turn a little in her direction.

What does the pyramid look like and what its levels are about?

Let's take a look at a well-known coffee brand as an example. Let's ask virtual users if they know what Jacobs is. Without clarifications and examples - just the name. To begin with, you can look at the drawing of the pyramid in order to understand which points are in priority when assessing the impact of the concept on society. And below you can read the form, where I will explain in detail what issues affect such an assessment and data collection.

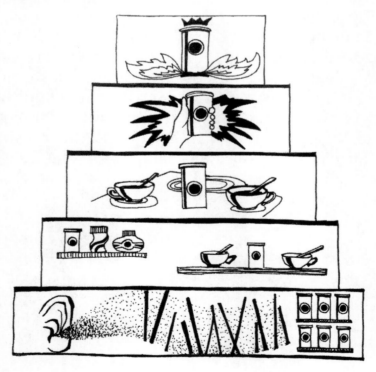

LEVEL	HUMAN RESPONSE	TREATMENT
1. I heard the name.	I have no idea what it is, but	This is the first contact with the brand.
I know that this exists.	I heard the name somewhere ...	The man just heard the name, but did not pay attention. The product is not particularly interesting to him.
2. I know what it is.	Yes, I know this is a coffee brand. I heard and maybe even tried it with friends a couple of times but I'm not sure, if it was this coffee. I drink very rarely, but if friends treat, then I can drink.	Person familiar with the activities of the brand, or even, perhaps, has already hit the target audience of the brand, but the brand isn't interesting enough to move customer to another level. It's just that this brand is one of many options.
3. I use it.	Certainly. I drink coffee regularly. But it doesn't matter to me which brand of coffee.	Here the person is already well acquainted with the brand. But it is still not fundamental to him, beloved and the one

	The drink helps me a lot to cheer up.	that takes the first place.
4. This is one of the best.	Well, if you already drink a drip coffee, then only Jacobs or Nescafe.	With a person's preferences at this level, everything is clear. Customer already sees for himself clear advantages of this brand and why he chooses it. Customer won't use other brands of this type and considers this brand to be one of the best.
5. Me and the brand are same blood.	I'll bite my throat for Jacobs. This is the coolest coffee in the world. Don't even try to prove to me that there is a better coffee.	A person will defend his preferences to the last, and even speak poorly of competitors, if necessary.

The fifth level is unattainable without randomness or, as it turned out, conceptology.

As practice shows, marketers cannot reach this level on purpose. As of today, there are no tools to predict ways to fully identify oneself with a brand, and to unambiguously push a person to firmly defend his position in relation to one specific brand. This is not affected by ads, creative efforts, or presentations in general. Conceptology is just that new tool that is able to predict the attraction of a huge audience almost immediately to the fifth level. Often without even going through the first three or even four stages. Because with the help of the actions and innovations that we offer to users, sooner or later, love with the brand will inevitably occur.

> We need to know this pyramid by heart, because by working with innovations, we will be able to determine which level we will adjust in the process of writing this or that strategy in the brief.

Any innovation guides concept users to the highest levels of the pyramid. It aims to influence people and their perception of the concept. For this, we not only understand the origins of different concepts, but also the mechanisms of implementation from development and improvement. As you can see, we also practice a lot. In fact, you have such a tool in your hands that helps to radically change attitudes towards brands, products and even people. If this, of course, is necessary.

Analyze your concept by obsessive development

At what level of the pyramid is the brand now?

_____

What is common in its main concept?

_____

_____

_____

How can you make it unusual?

_____

_____

_____

_____

§5.7 By destruction

This innovation is the hardest one, especially in performance. Its essence and trick is to force the concept to evolve through the complete destruction of its previous form. Sometimes the client simply does not want to understand and accept that a particular concept no longer works. Especially if this concept has been used by its professional niche for a long time. Although every concept dies after a while, our task is to replace this concept for our client so that he does not notice it or even himself participates in its replacement. Sometimes so that he further thinks that he is working with the same concept. After all, we were paid money for efficiency, and how to achieve this efficiency is already on our shoulders.

Each concept is, in fact, a system of views, a system of rebirth from one state to another. This system works as long as it becomes ineffective. Then we start looking for new solutions to the problem that has arisen, trying to improve the former state of this concept. In this case, we resort to innovation, which is the only solution to bring efficiency to the old concept.

Since the concept is always in motion, it begins to accumulate unnecessary experience that is largely unused and becomes even more ineffective. Eventually, a solution that works well becomes incapacitated, and a path of destruction begins. The stronger the destruction, the more this or that element of the foundation is checked for strength. But when this "beautiful" process of destruction is over, takeoff begins.

What should be said to a client who still orders articles in paper news newspapers, advertises his establishment with stickers on the fence, calls from his home phone, answers customers by mail, sells video tutorials on discs, packs products in plastic bags, is afraid to advertise his products through people with a different skin color ... And is a businessman lagging behind the modern rhythm?

But actually: "You don't have to wait for the natural destruction of the concept. If a particular business or organization is solving an important problem, the lessons learned should be processed regularly. If it has not been processed for a long time, other players appear on the market who process both their own experience and the experience of others. Then they enter the market with more interesting solutions than yours, and someone else will benefit from solving the problem. Therefore, we can not only follow the market trends, but also resort to self-destruction of this concept until its destruction happened by itself, without your participation. And instead of adjusting to market demands, we can already predict today whether the concept will be destroyed in the near future. If not, how can it be modified. This must be done before others begin to predict it!"

Innovation through the complete destruction of the current concept happens completely organically. It takes time, it is dictated by the era and this is a logical step in the evolution of this concept. It collapses by itself. Sometimes this organic process takes too long. Let's say like a book. Then it is necessary to stimulate such destruction.

Yes, an experienced concept scientist can suggest several options for the development of a concept. Including many other innovations. But when he sees that the only reasonable way out is to destroy the old form - so be it.

What does this look like in an example?

When the whole world moved on horseback, the profession of the one who was engaged in the "uniform" of horses was very important. That is, a person who completely made equipment for horses: a saddle, stirrups, special bridles and other assortment of effective decor. At that time, this profession was highly valued, and there were companies offering related services. Even with the advent of automobiles, companies with horse clothing services continued to operate. But some closed, while others did not. At the end of the 19th century, machines were made different from what they are now. They used a lot of wood and also required "equipment". For example, the same seats. They were made terribly uncomfortable and hard. Companies that used to equip horses have begun to create special seat covers, arm, handlebar, body and interior designs.

As Ford said: "You need to make money not by selling cars, but by repairing them." In fact, everyone adhered to this, in terms of car design. Therefore, smart guys quickly retrained into those who can make cars inside and out more beautiful and, most importantly, more convenient. In fact, they made the same saddles, but now in the format of car seats, and the same rubber harnesses, but in the form of brake pads and other elements.

The machine itself is an icon of conceptual innovation through destruction. There are gas lamps on the first cars, and electric on new ones, wooden wheels on the first, and tires on the next. As a result, even on carts with horses, they began to install wheels with tires. There are also curtains for glass instead of an umbrella in hand. And sprayers instead of wiping glass by hand. And yet another thing that manufacturers are proud of every year at technology exhibitions.

Moving from silent to voice-over

Remember the funny old black-and-white films where the characters moved strangely and then later films with text appeared? Until a certain moment, the movie was just like that. Later, a new format appeared, when the characters began to speak in a voice without decoding by text. It was no longer necessary to do musical accompaniment in the cinema, inviting a pianist to every screening of the film.

This is not just a natural enhancement of the concept, or even a collaboration. This is the real death of one era and the beginning of another. This is one of the most successful examples of innovation through destruction. It has been a reality for decades. This has resulted in the loss of some jobs and the acquisition of new ones. There are new emotions for the viewer and dozens of new or improved concepts that were born thanks to this decision.

Essentially, what happened? Gone is the era of silent films - and the era of films has arrived, where characters speak with their own voices. Where you can see not only their emotions, but also the intonation, the strength of the sound and the intensity and the timbre of the voice. Where the atmosphere of the film is supported by music and sound design.

All this is because a film was added to the visual film with pictures, which is capable of recording and producing sounds. One decision to improve a concept destroyed dozens and even hundreds of concepts that had been successfully operating before.

There are many examples of innovation by destruction

This same disruptive innovation occurred with slot machines in cafes and eateries when home game consoles were introduced. Before that, people went to special establishments to shoot primitive tanks or pick flat fruits. Moreover, these machines were really big. It was not only expensive to put it in your house, but in principle it was impossible. And what then? Who found those times, remembering how every child, and even adults, dreamed of becoming the owner of the console?

He was ready to leave all the slot machines in the world to stay at home, play with friends and not go out anywhere. There is nothing wrong with destroying the old and creating a new one on its basis. This is progress.

Destruction is not always complete

A USB or hard drive where e-books can be loaded hasn't destroyed the concept of a paper book. News portals have not killed fashion magazines. Even YouTube hasn't completely eradicated the love of watching the news or even your favorite shows in the kitchen, preparing meals. Netflix or Hulu haven't discouraged going to theaters.

Moreover, photography did not destroy painting. It modified and motivated it to experiment but did not destroy it. The concept of painting has undergone natural innovation for the same reason that everything changes. This is the variability of the surrounding world, trends, tastes, generations - everything that happens by itself over time. Portraits and family photos used to be painted to order. The photograph took it for itself. They used to paint landscapes. But photography does it better, too. Now artists are doing what they cannot photograph. To evoke feelings that a photo cannot evoke. That is, the concept worked, and it got better. Painted portraits have moved to another level.

Now this is what satisfies our aesthetic and status needs. For example, everyone can hang their own photo in their home. It's inexpensive, and you can take a photo with a good phone. Hanging your own painted portrait is completely different. Although this will soon also become as accessible as possible to every person.

By destruction

Probably, in the future we will forever give up coffee in the form in which we drink it now. At the very least, the preparation of coffee will be greatly altered. It will be replaced by coffee sticks or coffee capsules, which will contain coffee liquid instead of coffee. It is she who will make it possible to regulate

By destruction

the amount of caffeine, and instead of 20 cups from the same volume of coffee liquid, you can make 200 cups or even more.

They will be more convenient in format than anything related to coffee now. But one thing's for sure - conceptologists are not just the first to learn about innovation, but become its authors. And, perhaps, you will become one of such experts when you finally understand all the intricacies of creating concept books and conceptology in general.

Work your concept put by destruction

What's unusual about your concept in the last 10 years?

_____

_____

_____

_____

_____

_____

_____

_____

What's unusual about the concept lately?

How can you further develop these improvements?

How is it possible to amplify these improvements to the point of absurdity?

§6 CONSUMER CONCEPTS

Now it is even difficult for each of us to imagine that once in the world there were not many familiar things. Spoons, mugs, watches, towels, rooftops, doors or windows, and even roads. We live in rather comfortable houses and apartments, travel the world on buses, trains, planes. We buy packaged food, stylish clothes, soft sheets and blankets in stores, but we rarely think about where everything came from.

People always needed something, to catch a mammoth or other dangerous prey, so primitive people spent a lot of resources. Therefore, they did not live long, died in large numbers, returned from hunting wounded and sick and could no longer benefit their tribe. The desire to survive and reproduce made primitive human brains think and give birth to ideas that would help literally in everything.

Naturally, at first these were very primitive ideas. For example, people have noticed that a stone is heavier and more durable than a tree, and it is much better to hunt with it. Therefore, the first hunting tools were invented. If you grind the stone, they can hit the prey more easily and faster. And if you also attach a stone to a stick, then you can hunt at a safer distance.

Primitiveness is the cradle for the emergence of ideas.

Primitiveness is the cradle for the emergence of ideas, so-called industrial concepts, or rather, their prototypes; all that now we love so much, appreciate, and cannot imagine living our life without. Almost any household item of modern man originates from more primitive times, born solely out of desire, necessity, and need. For example, when people had already figured out that it would be more efficient for the community to slaughter a mammoth and then to store the meat longer, where to collect herbs and roots, and save supplies for the winter, it became necessary to record information.

And so, the first rock paintings appeared. How did people make them? They made them naturally, using improvised means. They noticed that unburned logs from the fire left traces, so they tried to draw with coals. They saw that a scratch would appear when pressing a stone against another hard stone, and so they began to carve drawings with stones. They noticed that some berries leave dark traces that last a long time, and they began to draw with them, adding different colors to the drawings... Why did industrial concepts originate in primitive times? Because people needed simple devices to satisfy their basic primitive needs.

And then, all that was needed was a push.

Humankind has been growing and developing, leading to new needs arising. The time of inventors, industrial revolutions, great discoveries, and even wars gave humanity many industrial concepts, each time in a new version. The primitive acquired an innovative look and more advanced functionality. A stone wrapped around a stick turned into a shovel and it helped the emergence of the bow and arrow, the halberd, the crossbow, and the modern AK-47.

The charcoal used to write has evolved into ink quill pens, slate pencils, ballpoint pens, and modern 3D pens. This way, we can trace the history and different types of almost every object in the world, decompose it into components, functions, and understand the etymology of the subject we use.

This will be useful for us to analyze concepts and propose new solutions in the future. Industrial concepts are relatively the most straightforward and most understandable, so we start with them.

§6.1 What do the lotus and the modern umbrella have in common?

PARASOL (fr. parasol - lit. "against the sun") - an umbrella designed to protect against the sun, made of light fabrics, paper, and lace. It was a fashionable accessory for walking in the 18th-19th centuries. The first kind of umbrella familiar to us was called the parasol.

Everything starts with an idea

An idea arises when a problem needs to be solved. The umbrella is probably the best example of this. The first umbrellas were utterly different from what we have today. Even back in cave times, people faced a problem due to water falling from the sky, and the sun was so intense that it blinded the eyes. This prevented people from tracking prey and recognizing enemies in time. So they had to think about how to get rid of this pain. They first chose what was always available for protection - large leaves of plants, which protected from the sun rays, but were quickly getting dried and lost their shape.

The rain also created failure because the leaves could not withstand the water pressure and were torn at the worst moment. Later, people came to the idea that the leaves can be dried, connected, and made similar to modern tents. However, even such constructions did not solve the problem entirely because they were too bulky and not mobile. These were only the first attempts of humanity to fight the rain and the sun. They used anything that could help, which was long before the invention of fabrics.

By the way, there are good mentions of the first "umbrellas" in the legends of the peoples of the East: 4000 years ago, a teenager with a large lotus leaf on his head was walking in the rain. The drops rolled down the leaf, and he did not get wet.

More than 3,000 years ago, a Chinese commander took soldiers to battle. It was really hot. Passing through a beautiful lotus pond, the soldiers plucked large leaves to hide under them from the hot sun.

The humankind kept growing but the problem didn't go away
Many scientists consider Egypt, China, and India to be the birthplace of the umbrella. Later, this item from the east came to Ancient Greece. Among other things, umbrellas have always been a symbol of power. Only emperors, pharaohs, and their relatives had them in China and Egypt, and these umbrellas were huge - 1.5 meters high and weighing about 2 kg. They were protected only from the sun. People started to improve umbrellas. To make the cloth, they used paper, which was soaked with special substances. The spokes were made from bamboo and reed, and the frame was made from bird feathers or palm leaves. The first mention of the umbrella in the form familiar to us is in the 10th-11th centuries, and it was only available to the rich. At that time, a pallor complexion was in vogue because it was believed that affluent people did not work in the sun since they were pale. Only poor people were demanding physical labor in the fields.

How umbrellas got to Europe

In the 14th century, umbrellas from China reached Europe. The women of France and Holland were especially fond of them. Umbrellas were graceful and elegant, made of lace and fine fabrics, and carrying them over a lady was considered a special honor for the servants. French women were fashionable with their various umbrellas, took them on walks, showing off who had the best umbrella. The richer the family was, the more umbrellas that women had.

Even Queen Marie Antoinette had her own collection of umbrellas. There was even a new position "honorary umbrella holder" in the palace. The Parisians offered special services where people could ask to be escorted home under an umbrella in sunny weather.

Until May 1715, the parasol was a one-piece frame structure (weighing about 9 lb) and stretched fabric. But in May 1715, everything changed. Parisian craftsmen invented a folding umbrella. This accessory became more convenient to carry; it became lighter, but, as before, it did not protect against rain.

The birth of the folding umbrella
In 1852, John Gedge told the world about umbrellas that could fold and open independently. The umbrella was no longer a solid and inflexible design. This invention was a real boom because an umbrella could be folded, taking up much less space.

Many people realized then that an umbrella could be used as a beautiful accessory for the sun and as a practical thing to protect from rain. The breakthrough occurred in the 1920s,thanks to the German Hans Haupt. He invented the first telescopic folding umbrella and founded the Knirps company in Berlin, beginning a real revolution

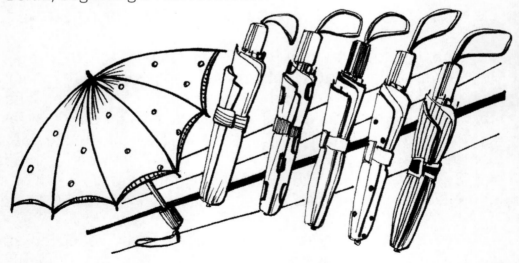

In the 1950s, umbrellas were no longer a fashion accessory. Instead, they began to perform only a practical function - to protect from rain. In addition, in the 1960s, nylon fabrics were introduced into production, allowing people to make umbrellas of different colors and patterns. They were also more durable and protected well from the rain.

The 21st century - the time of new developments

Every year, inventors try to make umbrellas even better and more functional. They came up with an umbrella in the USA with a built-in receiver connected to a weather station. If rain is expected, the light on the handle lights up. The brighter the light is, the stronger the precipitation will be.

The University of Tokyo is preparing to introduce the Internet umbrella to the world, where a satellite system will provide wireless Internet. This umbrella can be used as a screen to watch their favorite series or read valuable articles while walking in the rain.

A modern umbrella is similar to its predecessors when it comes to functionality. It has been evolving from an ordinary large leaf or palm branch into a compact accessory in such a way as to meet the needs of its time. Now let's move on to the form, to analyze this concept according to all the parameters you have already learned in this book. Prepare pens and notepads or notebooks to fill out concept books for various concepts together with me.

Brief project description

However, later people made them smaller. Umbrellas were used primarily for protection from the sun during the time of the first civilizations and later.

During ancient times and the Middle Ages, an umbrella was exclusively a ladies' accessory. For a long time, umbrellas did not fold up and only served as protection from the sun.

But then people came up with the idea to use them also as protection from the rain. This pushed to make the fabric stronger and the base of the umbrella more resistant to damage. There were different fabrics for umbrellas; people came up with folding umbrellas, completely transparent umbrellas, couture umbrellas, long, short, folding, cane-shaped, etc.

Now there is even a prototype of an umbrella, which we have something like a built-in monitor, so that while walking in the rain, you can watch the news or read.

Oral manifestation

This thing protects us from bad weather, such as heat or rain, not necessarily on a stick or with a cloth. The purpose of the umbrella concept is to protect us from bad weather or

Oral manifestation

unwanted influences. In addition, the oral manifestation of the umbrella is decorative to refract light or make it softer. Jugglers also use it as a tool and much more.

Actual manifestation

We talk about what we have now. This is a fabric stretched over rods, and this whole design is on a stick. It can also be a hat with rods, umbrellas on the beaches, etc. Photographers use umbrellas for their work. In the circus, umbrellas are used as juggling tools. There are even cocktail umbrellas. All these are just examples of actual forms of manifestation of this concept. You can add your thoughts as well.

Author's analysis

We talked about a short story and analyzed what happened. It's time to fantasize about what will be and what already is. The author's analysis is always about needs. Everything that we can invent and fantasize, one way or another, comes out of what people need.

- For example, there is a whole "culture" of exchanging umbrellas in Japan. In the cities, there are racks where everyone can take an umbrella if they forget theirs or leave it in the same rack - convenient Umbrella-sharing. People are reluctant to carry umbrellas with them. Why would they, if there is such a convenient option?

- Some umbrellas can often be seen right on people's heads. Such a great symbiosis of an umbrella and a hat. Comfortable and fun.

Author's analysis

• What else can it be? For example, the market has a rather poor color palette of umbrellas. So, it would be nice to release an umbrella of 10 shades into production. Also, nude shades are trendy right now, so why not make a stylish umbrella in 20 nude shades?

• Previously, an umbrella protected us only from rain and sun. But what if it will protect from other bad weather conditions in the future? For example, excessive humidity or gusts of wind. Global warming is coming; why not become a new trend? The weather certainly isn't behaving the same way it did 50 years ago.

Ways of concept development

Now we can definitely say that the umbrella should no longer look like an umbrella. It can be a dome covering a person or a group of people or a tunnel with a comfortable climate inside and rain outside. It can be a portable screen of thin fiber that broadcasts YouTube videos, music, or favorite movies.

Vulnerable points of the concept

Basically, everything that we do not like in the concept will give us a push to improve it. What do we not like about an umbrella? Why don't people use an umbrella? What stops people from buying an umbrella?

1. Its particular shape
2. Limited in colors
3. It takes up a lot of space

Vulnerable points of the concept

4. The very name "umbrella"
5. Umbrella does not fold
6. It pinches your fingers when you close/open it
7. Breaks and bends the other way
8. The umbrella flies away
9. It is difficult for people to constantly carry an umbrella with them since it takes lots of space in a bag.
10. It doesn't fully protect from rain.
11. When you close the umbrella, drops fall from it; everything around becomes wet.

Exploring paradigms

Let's analyze different types of people who encounter this concept and interact in various ways to understand how they look at the concept from their positions. Let's try to justify or condemn each side or find an average vision - a neutral position. **Let's divide each type of the studied subject into two points:** "Who?" and "What does it do?"

Who: Entrepreneur
What does he do: He looks for ways to use the umbrella with new features, a new look. To arouse consumer interest, use this umbrella as often as possible.

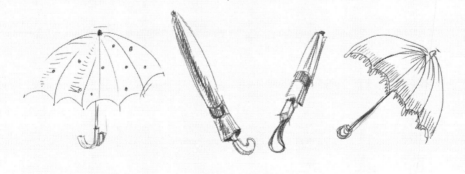

An umbrella is a thing that can be sold in bad weather. So, he is looking for opportunities to sell this product with a lot of sun or precipitation. He can create a special application that will remind people that they forgot their umbrella if they move more than 15 yards away from it (phone notification) and sell it together with an umbrella.

Who: consumer
What does he do: He looks for the product that will provide the best protection against bad weather, which will last long. Perhaps he will be looking for a product that turns the sun into coolness or absorbs (or does not absorb) moisture. He looks for a product that will not irritate him, will suit his style, mood, and character. Lightweight and not bulky.

Who: politicians
What do they do: They will think about making as many umbrella points as possible available where needed. First, of course, making it part of his election campaign, showing mercy and kindness towards his electorate, so that people know exactly who gave them this benefit, which specific politician.

Who: modern child

What does it do: For a modern child, an umbrella is not about the need to hide from the bad weather. Instead, he likes to jump in the puddles in the rain.

It's more about style, imitation, implementation. "Be like someone," "Imitate someone." It's something interesting, exciting. For example, Mary Poppins, an umbrella can make you fly away. Swimming in the puddles, why not?

Neural histories

Let's take related concepts and combine them.
Umbrella. Smartwatch. Space. Park.

Umbrella and smartwatch

You can build a voice assistant into the umbrella, make a watch face with a touchpad, put all this in the handle of an umbrella, play different music while walking in the rain.

Umbrella and park
Transform an umbrella into a format of a vast transparent tunnel through which you can walk and admire the flowing drops so that this tunnel protects all people from moisture or direct sunlight, with inside climate control.

Umbrella and Space:
An umbrella that can recognize the star map and display the starry sky on the umbrella's inner screen, even during the day time.

By the reinforcement of negativity

Let's suppose someone made a documentary about how a certain cartel is smuggling drugs, hiding them in the handles of umbrellas (or in other parts). This film shows how the

By the reinforcement of negativity

process took place, where they were hidden, and how much negativity and trouble it caused.

It shows how those who transported these umbrellas suffered and did not know what was in their umbrella. They would buy simple umbrellas, which were randomly mixed up in such a way that people did not even suspect that they were carrying drugs with them in rainy weather. Afterwards, the cartel representatives came looking for these umbrellas by all the means at their disposal.

Sometimes they managed to do everything quietly, but sometimes there were victims. People were afraid to buy umbrellas because they might contain drugs. But then the umbrella company investigated the case and made their own film that refuted the entire story.

They found out that nothing really happened. It was simply the idea of one filmmaker who wanted to bring the public to the problem of drugs in the world. The film went too far, but it brought fame to the director, although he had to help restore the good name of umbrella manufacturers.

After that, the people, on the contrary, wanted to buy more and more of these umbrellas because it made them feel involved in history and famous.

By rejecting bullying

We will analyze two audiences that argue with each other. They fiercely defend their point of view, which is diametrically opposed. We could talk about those who hate the use of

By rejecting bullying

umbrellas and those who cannot live without them, as well as about those who protect themselves from the rain with an umbrella, and those who use a raincoat to do so.

Arguments of umbrella fans	Arguments of raincoats' fans
An umbrella is stylish, beautiful, and not as ugly as a raincoat.	The raincoat is compact and does not take up much space. It is not heavy, unlike a bulky umbrella
There are many umbrellas of different shapes and colors, while raincoats are always the same.	Designers have been creating stylish and cool raincoats for a long time.
An umbrella has already been invented that has a built-in screen for broadcasting different videos.	The raincoat has special waterproofing openings with moisture protection for carrying your favorite gadget.
The umbrella may take different forms in the future, but the raincoat remains the same.	A raincoat can be transformed into a closet item, and an umbrella will always be an accessory.

By rejecting bullying

You can find fans of cane umbrellas and those who like short folding ones. Some will argue that a cane is stylish, convenient, elegant, comfortable. Others will look at them as fools, and prove otherwise. They will be arguing that only short folding umbrellas can have significant advantages.

For example, they are compact and lightweight. Or some wear them as a tribute to tradition, while others simply want it not to occupy their hands or even be hovering over their heads.

The bottom line is that you unnaturally create a natural excuse for arguments. Then you show that both sides are right. That's how consumers will fall in love with your brand.

By the miracle effect

The main trigger is an appeal to the inner child. That is, working with the effect of a miracle, we touch the strings that evoke children's sincere delight. We do something that will cause such delight which only real magic can do this way. So, an umbrella. At first, we look for what annoys us, but we have long been accustomed to this.

Main question	Irritants
What annoys us, but we are used to have it?	- An umbrella can only be carried by a stick.

By the miracle effect

What annoys us, but we are used to have it?	- Even with the best umbrella, feet are still splashed and get wet. - All umbrellas are mostly gloomy shades. - An umbrella takes up a lot of space. - You can forget your umbrella easily. - One hand is constantly busy. - If it is an umbrella used against the sun, it does not protect you from all the rays, and you can get burned.

What can we do in this case to create something extraordinary?

1. Create an umbrella backpack that does not need to be carried in your hands. Instead, simply an umbrella will be taken out of the backpack itself and, like an awning, stretched over your head.

2. Create an umbrella that will change colors depending on the settings.

3. Subscribe to an umbrella or one-time insurance. If you lose or forget your umbrella, the company will send you a new one or give you a good discount on purchasing the next one.

By the miracle effect

4. Make an umbrella that will follow the sun and protect (such a "sunflower effect") from the ray, or recycle the rays and modify them into useful ones.

5. Sell a bracelet along with an umbrella, which will always remind you that you forgot the umbrella.

By compulsive reputation improvement

So, we know that something is waiting for us ahead that will definitely shake the reputation of our brand /concept. We are aware of this, and our task is to do something that levels the blow from the coming event. Where is the point of conflict with the umbrella?

For example, when creating a completely innovative umbrella, many people will say that it does not even look like an umbrella. The new umbrella does not have main characteristics, which means that it cannot be called an umbrella either.

To improve the reputation, we need to know what advantages we have now.

What are the advantages of the concept now?	How to use them when your reputation has been hit?
Rain and sun protection	Any innovation will fulfill this function and make it better. For example, climate control inside the dome. It protects us from the rain

	By compulsive reputation improvement
	and creates comfortable conditions inside.
Image	This is something that you can carry with you into any concept. It is enough to make the innovation that everyone wants.

This is a life hack: we can start not from destroying our own reputation. **To improve the reputation of your concept, you can use the good reputation of another concept.**

We can build on the fact that another concept has done it before, but we made it better. For example, to show that the umbrella did not always look the way it does now, which means that it may well change in the future. In other words, earlier, it was generally a huge leaf and then a huge iron tent, and now it is on a stick, and then it may become something above the head. And this is now NORMAL. In the same way, a phone with buttons turned into a touch phone.

By collaboration

We create a symbiosis of a concept with another concept, improving its functionality, but without categorizing anything. An example would be to create an umbrella made only for a child who is carried in a sling and then collaborate with sling manufacturers.

To create an umbrella and a backpack, create an umbrella

By collaboration

that slips out of the backpack and takes up little space or additionally, collaborate an umbrella with other eco-projects. One example would be to create an umbrella from eco-materials that absorb solar energy to store it so that they can charge small gadgets.

There can be many such collaborations. It can even be a collaboration of one umbrella with another. For example, one that folds and one that is in the form of a cane.

People can create an umbrella in the form of a cane, but if desired, it can be easily folded and hidden in a bag. At the same time, the quality of the cane is no less than in those where it is solid and does not fold.

By destruction

We see that someone uses the concept better than us or just like us. We take everything we know about the concept and improve it comprehensively. It's an umbrella that no longer looks like its first form. From the very beginning, we continue to improve it to the point of absurdity. For instance, there can be umbrellas over entire cities, covering the cities when it rains.

There can also be smart cars in the form of small transparent balls that protect us from the rain, covering us with a kind of cloak within a small space. There could be a comfortable chair and climate control inside, so no more worrying about forgetting your umbrella at home. There will be parking lots throughout the city, and you can always rent one for yourself to get there.

It can also be a spray umbrella that you spray on yourself,

> **By destruction**
>
> and it covers you with a thin film that repels drops without making you feel uncomfortable.

   This is just an approximate analysis of one concept. I am sure that even in the process of reading, you had your own ideas, and you wrote them down somewhere. But, even if not, do it urgently before you lose the thought. Because one of the most important tasks of conceptology concerning its experts is to start the mechanisms in the brain so that they can generate new ones. Indeed, the new often combines what we already know and what we can do. Only now are we looking at this set from a completely different, often even unexpected angle, and the result is stunning.

§6.2 Walking over mud and podium? The invention of the heel

Heel - a hard shoe bottom on the sole of the shoe, a detail in the form of a vertical stand that raises the heel above the toe level. Heels on shoes appeared in the late Middle Ages. In Russian, the word "heel" was first noted in written sources in 1509, probably borrowed from the Turkic kabluk, which comes from the Arab. kab — "heel".

English heel comes from Old English hēla; related to Old Norse hæll, Old Frisian. hêl. "In the sense of a rounded rear part of the foot".

Spanish Tacón (the part of the sole of the shoe that covers the heel). Tacón comes from the same Germanic root that gave the word Taca, used in the mining industry ("forge crucible plate") and in the sense of "a part of an object with a color that is distinctive from the common".

The first "trendy people" were farmers?

Now we are used to seeing women in high-heeled shoes and men in formal shoes with small heels. But what if this is not just a tribute to tradition - the roots of heels are much deeper, and its very history is caused not by fashion trends at all but by a rather prosaic necessity? Let's analyze where the idea of wearing high-heeled shoes came from, and what was its purpose?

In ancient times, shoes were an expensive pleasure. For example, in Egypt, they were made from papyrus and could only be worn by the pharaoh and those whom he honored. It is not surprising that the price of such shoes was equal to that of the royal standing. Herodotus wrote that one pair of the pharaoh's sandals was the same as the budget of an average city for a year. The first heel prototype in Egypt did not appear among the pharaohs at all, but among ordinary workers who worked in the field. The tillers built something similar to a modern heel and fastened it to their feet to make it more convenient for them to walk on a loose surface and create a grip. Butchers also used a similar method to rise above the floor and not step in the blood and waste of animals. At the same time, in the East, shoes "with heels" were worn mainly in ... baths and saunas so that the hot floor did not burn the feet because the temperature of the stone surface was very high.

Theatrical fashion and its followers

In Greece, heels were worn to the theaters. However, it was not those who came to see the performances and show off their outfits, but the actors themselves. When playing the ancient Greek gods, the actors always wore high heels. It was believed that the gods should always be above mere mortals - "float in the air," towering above the rest. Therefore, the actors had to go through such efforts to create the illusion. However, this idea was adopted by the Greek courtesans in their way. It is believed that they came up with the first prototypes of the stiletto heel. They asked cobblers to tamp shoes with nails to leave a barely visible mark that seemed to say: "Follow me." In Japanese theaters, men also wore shoes with heels. At first, it was only available to actors, and then subsequently the nobility adopted this style.

By the way, Japanese geishas also wore shoes with heels to protect and not stain their elegant yukata and kimono in the mud. Venetian women of the 15th century wore large platforms (sometimes more than 8 inches high). They were called tsokkoli (or hooves), because when they walked in them, one could hear the sound "tsok-tsok".

And in Europe, everything was different: the trouble that prompted the invention of heels.

It is an unpleasant fact that European history and architecture have a lack of a well-thought-out sewage system. There was literally nowhere to put sewage in cities, and the streets simply sank into the foul mess. Even kings frequently moved out of their castles because of uncleanness. So for cities' residents to be able to walk normally and not get dirty in the mud, they came up with shoes with heels - sabos. They looked like ordinary wooden soles with leather straps attached. People wore them over ordinary shoes. The fact that shoes with heels began to spread was also influenced by military innovation. This story goes back to the Baroque era. At that time, horse riders wore similar shoes: a heel helped the foot keep the legs in the stirrup and not slip out. Because infantry soldiers had to walk far and long distances, they needed special shoes. This is how high-quality boots appeared from craftsmen who made them with thick soles and attached heels. By the way, the shoes of those times were completely uncomfortable. Because no one was thinking about convenience, people sewed so that no one distinguished the right and left feet.

The trouble of complexes of nobility and kings

It was a-woman who came up with the idea to show that the heel can also be used for aesthetic purposes. Catherine de Medici, Queen of France, dressed up for her wedding in shoes with 2 inch heels. The other women at a royal court liked her idea so much that eventually the whole world wore such shoes.

Mary Tudor was an arrogant lady and did not want to be on the same level as anyone else. It was for her that they came up with a heel in the form of a cone, which is similar to the one that we use today.

Louis XVI was sad because of his height. As a result, this monarch's inferiority complex boosted the development of high-heeled shoes. However, the king did not know the proper measurements, and as a result, there were shoes with heels up to 24 inches in his collection.

The rough heel was before the invention of stockings

In the 17th century, knitted stockings were fashionable, leading to a boom in shoes that resemble modern shoes. In 1680, the year was challenging for the ladies because it was fashionable to wear super high and thin shoes. Walking in such shoes was possible only by leaning on a cane. Men also did not want to be left behind, and the fashion for heels seized everyone, to the point that the kings personally approved the height of heels by class.

High - for the nobility
Low - for everyone else
Red - to distinguish the nobility

Heels. Revolutions. Twentieth century. Stiletto

The French Revolution of 1789 took away women's heels for almost 50 years. The shoes of that time were more like ballet shoes because then Europe wanted a simple life and didn't welcome complicated outfits. In addition, heels were considered unhealthy because their shape was uncomfortable and distorted the foot. But the twentieth century was just a paradise for shoemakers.

During this time, so many styles and types of shoes were invented as never before. In addition, the natural rubber helped make the soles waterproof and the shoes more practical.

While new materials were used to produce shoes (metal, plastic, rubber, etc.), shoemakers came up with a stiletto. Thanks to aluminum and the injection molding method, it was possible to couple the metal base and plastic. That gave the world stiletto-heeled shoes.

Such shoes were originally called stilettos - because of the resemblance to a sharpened, pointed, thin dagger. The first "stilettos" were produced in Italy in the early 50s of the twentieth century. From that moment on, these shoes have firmly made their way into the collections of world designers. In 1953, Roger Vivier created the most beautiful heeled shoes with precious stones. This was done for the coronation of an English princess. This model caused the "boom" and strengthened the position of the "stiletto".

Heels features in the light of dance life

Special attention deserves shoes for Latin American and ballroom dances, always performed on the heels. But if you look closely, the heels are different, and there are several nuances here. Dancing heels may not seem the prettiest, but this is not about beauty, rather about convenience. The shape and location of the heels are developed in special research institutes. Each new model goes through something like "testing in the fields." This is done by dancing couples and a certain committee, including distributors, coaches, and dancers. The shape of such heels is different:

The Cuban heel
is quite wide and beveled towards the bottom.

The Latin heel
is just wide without beveled edges.

Block heel
wide rectangular shape, which is very stable.

The flared heel
comes in different heights and thicknesses. It is wide at the base and slightly narrowed towards the bottom.

Это небольшой перечень основных форм танцевальных каблуков. Главное правило — они должны быть суперустойчивыми, даже те, которые узкие и косвенно напоминают шпильку. Чаще всего используют форму с широкой набойкой — для лучшего сцепления с паркетом и большего баланса движений танцовщицы. А вот высота каблука может быть разной для всех форм.

Heels – an open space for designers' creativity

It's great that now we no longer need to rise above the waste or cling to something on the surface of the earth so as not to load it. The heel is now an exclusively aesthetic story. Designers let their imagination guide them and develop the most unexpected and intricate forms. There are 7 basic heel shapes:

Vienna heel
is a short (up to .6 inches), almost imperceptible. Often shoes with such a heel are not considered "heeled."

The brick heel i
s rectangular, height up to 1.5 inches, and looks like a brick.

The wedge-shaped heel
resembles the shape of a wedge, narrows from the base to the bottom. It can be any height.

Cowboy heel
is low square with a beveled part at the back. Mostly, it can be found in the design of "Cossacks" and boots with a wide top.

The glass heel
resembles the stem of a glass. It is wide at the base and gradually tapering towards the bottom. Usually, such a heel is not higher than 2 inches.

Heel-column is pretty stable.
Straight heel, quite high (from 2 inches).

The stiletto
is somewhat reminiscent of the shape of either a needle or a nail. Despite the relative thinness (diameter up to .8 inches), this heel is quite stable.

The stacked heel
is also called puff. It consists of flicks layers (they are made from an intermediate material, mainly leather or cardboard).

Figured heel
well, here the word speaks for itself. This heel can be of the most bizarre forms: consist entirely of rhombuses, resemble animal hooves, be one solid figure (for example, a heart). Nevertheless, it is the designers' favorite form of the heel.

A wedge heel
is also a kind of design solution. There are two opinions that a wedge is a sole form since it is solid and one-piece. The second one is that the wedge heel is a modified heel format since the idea of elevation at the heel is still borrowed here.

Shoes with wedge heels are trendy among those who cannot wear heels for a long time but want to visually appear taller. Moreover, designers use wedges often. This starts from casual shoes such as boots or sandals, ending with sports shoes: kids, running shoes, and sneakers.

The task for developing the habit of collecting data:
See what each type of heel looks like online and find them in our pictures above.

Brief project description

In prehistoric times, people did not immediately begin to wear shoes, but only when they realized they needed to protect their legs from sharp stones. This was common during long migrations or hunting, but there was no talk of heels.

Oral manifestation

It's the thing that makes people taller, helps them rise above the surface. Protects us from the dirt on the ground. It helps us look slimmer and more confident. It helps dancers to make different sounds while dancing.

Actual manifestation

We talk about what we have now. This is a specific thing made of plastic or polyethylene, or other materials, which is often used as an additional body part of the footwear. Often it does not perform a solely practical function, but a decorative one. It can be of various shapes and types.

Author's analysis

We talked about a short story and analyzed what happened. Finally, it's time to fantasize about what will be and what already is. The author's analysis is always about needs. Everything that we can invent and fantasize here, one way or another, comes out of what people need.

- There are heels that you can take off and walk just on a flat sole.

- There are heels with special holes to put something in there.

- Some heels help the foot hold on to the surface better so that feet do not slip and help people not to fall.

Author's analysis

- There are exclusively decorative heels used to shock the audience at concerts.

- Some heels originally repeat ordinary household items, such as eggs or towers.

Ways of concept development

What if the heel doesn't look like a heel? What could it be? When people came up with connecting the sole with the heel and making a solid wedge platform, it was already a step towards changing this concept. To understand the development of the concept, you need to grasp the main thing: what purpose does it fulfill? In the future, the heel may well turn into a force field that will lift a person above the ground to the desired level. It can be regulated and controlled.

Exploring paradigms

Let's take different types of people who encounter this concept and interact in various ways to understand how these people look at the concept from their own position. We are trying to justify each side, condemn or strengthen it, and find a unifying vision.

Who?	What are they doing?
Those who wear heels always and everywhere	They give 1001 arguments that wearing shoes with heels is beautiful.

Exploring paradigms

	It makes a person taller and more elegant. Legs in heels look slimmer and sexier. Heels help to improve and lengthen the figure visually. They say that heels are art and that you need to wear them, and not everyone can do this.
Someone who prefers low heels	Talks about heels are not good because they deform the foot, and due to the uneven distribution of body weight, health problems often arise. That beauty is not worth it to suffer from pain in the legs. Walking in high heels is no longer necessary because the streets are not as dirty as in the Middle Ages, and the heel no longer fulfills its original function.

Neural stories

Heel and car

People can come up with special holes in the car where the heels can be kept. It is very convenient to remove the heel, especially for women who drive and experience discomfort from the way the foot is on the pedal. It'd be a great move to place that extra shelf under the seat and keep your heels there. This would be especially convenient if you combine history with a collection of different removed heels and the moment when they can be hidden in the car.

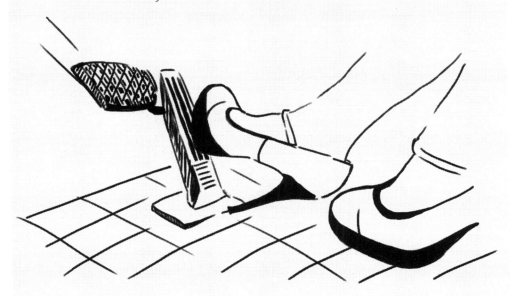

For example, you go to work in sandals with low wedges, and then you add to them a new heel - and you are already on a stiletto.

Heel and flash drive/media

You often need to carry a bunch of information with you separately, and there is nowhere to store it. Even if you take a small flash drive to keep all the information, it can get lost.

Neural stories

But if you come up with a particular hole in the heel or make an information carrier there, you would not need to worry about where this flash drive is constantly located. Especially if you consider that you will still definitely be in shoes at work, and the heel in its shape will be the most convenient place to hide something there.

Vulnerable points of the concept

What do we not like about heels?

1. If you wear shoes with heels for a long time, your feet hurt.

2. You can experience not only discomfort but also may get varicose veins.

3. The heel cannot be unhooked from the shoe when you are tired of it.

4. Often the heels fall into the holes in the drain gratings and make holes in the ground. The dirt sticks to them, and it looks terrible.

5. Heels constantly need to be changed; they are easily worn out.

By rejecting bullying

We analyze two audiences that strongly disagree with each other. They fiercely defend each point of view, which is diametrically opposed to each other.

By rejecting bullying

Arguments of heels' fans	Arguments of heels' haters
Heels help you look slimmer, prettier, and sexier.	The sole without a heel is comfortable and practical
Wearing heels is valid, it helps to pump up the leg muscles.	Platform shoes are equally well suited for business and elegant collections.
Heels are a lifesaver for short people. It always adds + a few inches to height.	When you walk in shoes without a heel, your legs do not hurt, and no varicose veins.
There is a specific dress code, the element of which is heels. If you don't know how to wear them, they won't let you into a private club.	Shoes without heels are easier to store than shoes with heels.

The wars between those who wear heels and those who despise them may continue for a very long time. After all, some look no longer at aesthetics, while others look more at the consequences.

Someone chooses beauty at the expense of their own health, while someone, on the contrary, is ready to sacrifice

By rejecting bullying

their beauty to be healthy later. The main thing is to find where this confrontation can originate and create a new form of heels or further develop the old one.

The bottom line is that you create a unique reason for rejecting bullying. Then you show that both sides are correct, and their opinion is reasonable, so you will make consumers fall in love with your brand.

By collaboration

Let's talk about heels. What do many girls dream of since childhood? Of course, to try on a crystal shoe and feel like a real princess! What if we give them the opportunity? What if we come up with a shoe that appears all at once, along with heels? Or, like in a fairy tale, we will show the story of how a girl puts on only one shoe, gets the whole elegant bow, and hurries off to a date.

By collaboration

All these stories and the princess archetype can easily be used in advertising campaigns and promote various brands through them - from cars to clothes. You can perfectly connect innovation through collaboration, which designers often like to use. For example, in 2021, a whole collection of shoes came out with the most bizarre heels; there were scrambled eggs, nails, and even supporting hands. Sports shoes also once collaborated with heels, and there were sneakers with heels. However, fashion did not accept them since they looked more bizarre than attractive. If you combine the effect of a miracle and a collaboration, you can create a fashion show, where each designer presents their pair of shoes. As soon as the

By collaboration

model puts them on, they immediately put on some appropriate fabulous outfits. Isn't this a miracle or is it instead an extravaganza?

By destruction

How to destroy the concept of a heel? This is not possible in the near future. Firstly, because the heel is firmly established in fashion houses and is a part of many models of sports shoes, where there is actually a hidden heel, designed so that it correctly distributes weight on foot. Although this is not the heel to which we are accustomed at all, the world is changing, and perhaps in the future, people will no longer treat shoes like they do now. Perhaps we will develop a special anti-shock orthopedic coating for the legs.

Or even just insoles that will take the shape of the feet and then be shaped into shoes. Most likely, the heels will disappear and become history. Perhaps they will remain part of beautiful and fashionable images, but that will be a completely different story.

By obsessive development

This is a heel that no longer looks like its first shape. In the beginning, we improved it to the point of absurdity. For example, it can be that such a heel that makes a person taller and allows him to fly above the ground and not to get any stains on his shoes at all. It could be a folding heel that changes the height of the shoe or a heel that transforms into wheels and makes shoes something like rollers, which allows you to move around the city quickly.

Research concepts are some of the most interesting because they are born and exist together with ambitious thoughts, ideas to study something hidden and incomprehensible, inspiration for great achievements, and the desire of people to do unusual things, create great goals, build academies, improve various state institutions and open entire scientific areas. For example, turning the ordinary process of cooking into the art of cooking, or dreaming of finding out what is beyond the horizon of the seas, building a ship and going on an exploratory journey, or watching the stars for half a lifetime and record all the changes so that your descendants can one day fly there.

The main engine of research concepts is the idea to achieve more, learn something new, explore the incomprehensible, to get something that was not there before. For example, in all historical eras, music followed people. Can its structuring into a conservatory, orchestras, and even sound media be called research concepts? Why not?

Humanity's first musical attempts were related to security, and it was a way to convey information. But after all, we noticed that different sounds could be extracted from different natural materials of different intensities and grace of sound. How did we use it? It turned out that it was used not just for sound alerts.

Music rushed into the world of people like a whirlwind. The first musical instruments date back to the 3rd-2nd century. They were not created to satisfy some basic need but the result of a craving for beauty. Music and its changes have been studied extensively. Over the centuries, many musical instruments appeared - from beaters to strings, from a shepherd's horn to wind instruments.

Yes, the music itself has changed. Each era gave birth to new styles and trends. Music has become a tool for self-expression and a way to tell the world who you are. As a result, it grew into a whole industry divided into many branches and took on the world as a whole.

Now music is not just the sounds of balls and orchestras. This is practically the very accompaniment of life: it is everywhere and in different formats and in the ears of almost every person at any time of the day. How could we think that we would carry what we used to store in entire huge rooms and now in a small box? And this is far from the limit.

Research concepts are the engine of progress

The essence of research concepts is to transform simple needs into art and science. If you look at the big picture, research concepts can cover more than one area and involve more than one concept. For example, for the same cooking, these include industrial (kitchen utensils), economic (the ability to buy for money, deliver from a restaurant through a delivery service), and philosophical (myths and ideas about the emergence of different cuisines of the world).

These are research concepts that boost great discoveries. Thanks to constant discovery and invention, we get spaceships, electric cars, flying boards, high-speed trains, and more. This section will discuss those research concepts that seem to you to be the most understandable.

But in fact, the mission of this chapter is to dig deeper and find many unexpected insights that will help you on your way to becoming a professional conceptologist.

§7.1 Your delicious lunch through history: How did humans invent cooking?

Cooking food (lat. culīnāria "kitchen <craft>"; from culīna "kitchen") is a human activity for cooking, which includes a complex of technologies, equipment, and recipes.

Cooking is a set of methods for preparing various foods necessary for human life and health from minerals and products of plant and animal origin.

Once the meat accidentally fell into the fire

In cave times, people were not picky about their diet. Back then, there was no such thing as proper or improper nutrition. No one thought that food needed to be cooked and that some great food combinations made dishes incredibly tasty.

People survived at the expense of the prey that predators did not finish, and nutritious plants and random discoveries they made by tasting everything in a row. At the dawn of humanity, they sometimes had to determine the edibility of particular food at the cost of their lives. People ate simply because they had to survive; there was no question of food's taste and gourmet characteristics. All this lasted until the meat of a butchered mammoth accidentally fell into the fire one day. It was the only meat, and the cavemen realized they had a choice: either leave hungry after a long hunt or eat what they had. Nobody wanted to starve, so they had to eat. They were surprised - the meat turned out to be tastier.

So, a random event pushed the world to the birth of cooking. Further, this art kept on developing with each new generation. Primitive people roasted meat and fish on the fire, often used hot stones, and baked it. One of the popular dishes of the Stone Age was porridge made from different cereals, something like a modern polenta, grits, or banosh. Some people have crushed corn with other ingredients, wrapped in corn leaves, and stewed on fire. These dishes were prepared easily and were quite nourishing. Over time, people came up with the idea of growing herbs and adding them to food as seasonings to enrich the taste and prepare simple stews from various ingredients. Humanity very quickly realized that delicious food is more interesting and was constantly looking for ways to improve what nature gives.

Goddess who fed everyone

At different times and regions, people perceived food differently. Sometimes they praised it; sometimes, they considered it unworthy even to mention in conversation. Some people "lived to eat," while others "ate to live." For example, they ate simple food in ancient Sparta - cereals and stews, flatbreads, and water.

The cooks of Ancient Greece roasted a bull on a spit. Inside the bull, there was a roasted sheep, a goat in a sheep, a dove in a goat, and an olive in the dove. Despite your surprise, they did so to "be beautiful", and of course, for the creation of a new taste.

In ancient Greece and Rome, the goddess Kulina was very honored and was considered the patroness of tasty and healthy food. She was the nurturer of all the gods of Olympus and the tenth muse for the Hellenes. The Greek fashion for gourmet food was passed on to Ancient Rome, where a talented chef was the sign of a successful house; no wonder it was there that they began to teach culinary science.

The cooks recruited apprentices and taught them how to cook. They passed on the secrets of cooking only to the smartest; therefore, professionals who knew how to combine the right ingredients were very much appreciated in those days. At that time, there was a culture of food, which we now call the "Greek diet ."The bottom line is to divide the products into those you need to eat every day, week, and month.

In more detail, it looks like this:

Daily
- bread, pasta rice
- vegetables - from potatoes to eggplant
- fruit
- nuts, sunflower seeds, or melons
- olive oil - a few tablespoons
- yogurt and cheese
- 2 glasses of dry wine per day
- at least 8 glasses of pure water a day

Every week
- fish and seafood
- eggs
- some sweets.

Every month
- lean meat and poultry;
- red meat.

The Greeks have always loved to feast, eat delicious food, philosophize, and develop a new dish. Homer, Herodotus, Plutarch, Aristratus, and many other famous figures devoted entire chapters to describing food and drink. Later, the food cult spread to Rome, Egypt, and the emerging European countries. But in the East, modesty reigned, and food was rarely perceived as something special.

Often, many foods were not even cooked there and were eaten raw: some types of fish, fruits, nuts, herbs, and even mushrooms. But spices have always been highly valued there, giving a unique taste to different products. The Europeans made great efforts to procure these spices through the centuries.

There was no time to invent unique dishes

The Middle Ages slowed down the development of the art of cooking. The finest delicacies were available only to kings and nobles in many regions. Ordinary people ate what they farmed. They had no time to think of cooking as something lofty. They worked hard, so they saw food only as a way to get energy and not die.

Everything changed when the spicy flavor of the East entered Europe, cooking rose to a new level, and Italy became the center of the culinary renaissance at that time. As for French cuisine, the culinary history boosted under Louis XIV - that's when the "war" began among European chefs and culinary specialists.

In the 19th century, the first culinary school in England appeared after a great lull in this field. In France, cooking has been and remains very popular. Antoine Karem distinguished himself during the time of Napoleon. He knew the history of his profession, especially the ancient Roman cuisine, which was famous for its heavy and fatty dishes. In his notes, he pointed out the dependence of mood on food: "A writer knows how to enjoy good gastronomy. Poets love a good meal and praise ambrosia. The gloomy philosopher pays little attention to gastronomy."

An institution similar to the English one, divided into male and female groups, opened its doors in 1891 in the French capital. It was no longer just the transfer of primitive knowledge about baking bread or cooking game. It just became boring to eat food, and the audience demanded unique features. Each region began to develop these features in its own way.

National cuisines and their features

Changes in cooking depended on the characteristics of the country. For example, Greece was rich in vegetables, and the climate was favorable for growing various crops. In addition, the Greeks were very fond of different combinations of food. By the way, the Greeks came up with the prototype of borscht - a vegetable soup in meat broth with beets and cabbage. Bread and cereals were the leading food in the ancient world. They were used to make congee and porridges, such as masa, a mixture of flour, honey, salt, olive oil, water, turon, flour, grated cheese, and honey.

Many foods were sprinkled with barley flour before cooking. Beans and other leguminous plants were abundantly used. In the Bronze Age, many vegetables were already known and used. Sometimes lamb or beef was added to vegetable dishes, but the meat of domestic animals was expensive, and hunting trophies were widely used - the meat of wild animals and birds, which were abundant then.

The favorite national soups of the ancient Romans were a variety of cabbage soup and borscht - especially for them, a lot of cabbage and beets and onions were grown on agricultural estates.

However, the Spartans had a relatively meager diet. They did not raise food to a cult status and ate modestly, but enough to get energy for battles and campaigns. A known recipe for Spartan stew – Melanas – was a piece of meat or pig's feet, bull's blood, vinegar, salt, and sometimes lentils were added. Obviously, the taste of the stew must have been just terrible.

Talking about the features of European cuisine, each country was already famous for its own signature dish. These were crispy baguettes and croissants, onion pies and soups, and fluffy buns in France. England was known for its love of breakfasts, toast, cereals with various fillings (berries, nuts, jams, fruits, etc.).

Austria was famous for its pastries, schlipf krap dryers, schnitzels, and sausages. Hungarian - goulash, bigus and numerous dishes from mangalits (curly-haired pigs). Germany - various types of sausages, schnitzels and Sauerkraut.

Poland was famous for pies, various mushroom zhups and baked casseroles. Ukraine - dumplings, borscht, galushkas. Russia - pancakes, mushroom stews, cabbage soup, and all kinds of cooked wild meat. The list can go on. But here, it is essential to say that the national cuisine of any country was formed based on what products were easily accessible.

If we talk about eastern countries and those now called third world countries, then there is also one crucial nuance here. The dishes of this cuisine are mostly very spicy. Here, spices were originally used not so much for taste as for internal disinfection since unsanitary conditions and gastrointestinal diseases are common in these countries. As soon as the inhabitants of the East discovered the beneficial and antiseptic properties of spices, they immediately began to use them abundantly in cooking.

Cuisines and spices

Spices in cooking have a special place. In the past, entire expeditions were equipped for spices in India. They diversified the range of tastes of European cuisine so much that chefs from many countries fought for every gram of the wonderful powders. Famous travelers talked about the "fragrant" product, bringing new types of spicy plants from distant wanderings. Many of them left their notes.

The monk Cosmas Indikoples in the book "Christian Topography" (530) described some spices. Venetian traveler Marco Polo published a book describing his discoveries and wrote about spicy plants.

This book greatly influenced Christopher Columbus, arousing his interest in finding new sources of expensive goods - spices. The Portuguese Vasco da Gama expedition was also full of adventures: many of its members died, and the expedition lost its ship. The income from the sale of imported spices significantly exceeded the cost of the ship. In 1519, the Spanish Pacific expedition of Magellan was equipped.

Of the 265 sailors, only 18 and the flagship returned. Nevertheless, the value of the cargo of cloves they brought back exceeded the value of the lost ships. Subsequent travels, wars, trading expeditions contributed to an even greater spread of the "spicy" goods in all parts of the Earth. Gradually, people learned to use spices as a fragrant addition to the dish and as medicine with miraculous properties.

The world tried to get spices to make their dishes even tastier and more refined by hook or by crook. With the advent of spices, many familiar dishes have changed, creating a new round in the cooking development. The purpose of cooking was no longer the relief of common hunger, now cooking began to fulfill a new task - it became an art.

The War for Recognition, Molecules, and Stars

As we discussed earlier, the profession of a cook has been valued since ancient times. In the Middle Ages, cooks became so popular that entire culinary schools were opened, although cooking deliciously was only for the nobility, nobles, and the king. Simple cooks could not create exquisite and unusual dishes, as the aristocrats and the elite required.

A cook who could come up with something new himself was in demand and always held high esteem. Often, they became famous thanks to ingenuity and chance. There is one famous story about cheesecakes with raisins. One strict nobleman constantly ordered cheesecakes from the same cook. The recipe did not change; it was buns with sweet cottage cheese and a crispy crust. But once a cockroach fell into the cottage cheese, the cook, to get out in front of the nobleman and not lose his head, came up with the idea that this is a new recipe for cheesecakes with the addition of raisins. The nobleman did not begin to understand why raisins appeared in his cheesecake, but to the cook's delight, he appreciated this innovation and fell in love with cheesecakes with "raisins." Who knows how many popular dishes arose due to such a coincidence.

Over time, the cooks cooked for the nobles and opened their own establishments where several people could eat at once. So the struggle began to attract as many customers as possible. The menu was updated, the variety of dishes was amazing, and the approach to cooking was surprising.

More attention was paid to the design of dishes and their decorations. There were beautiful patterns of fruits and herbs on plates, decoration and serving techniques, devices for decorating various dishes, and even shows. At first, these shows were rather primitive, but over time, even shocking. For example, they could cut the ribs right on the table, and then go cook them, or serve more BBQ wings blazin with multi-colored fire; or bring ice cream on a tray that smokes from liquid nitrogen, or surprise serving, where a wooden block is used instead of the usual plate, has become increasingly popular.

Further on, cooking developed in gourmenism. Both chefs and customers were looking for a new taste and a new form. This is how molecular cuisine was born. Now the chefs were able to deceive your eyesight and confuse your brain, offering to eat something that everyone is used to, but in a completely unexpected way, such as sushi in the form of a cake or pate disguised as tangerines.

The culinary masters played with shapes and textures. They figured out how to make the middle of a whole dish pour out, for example, in poached eggs or chocolate fondant. They came up with different ways of slicing vegetables and meat to show a new taste, like in tartare or carpaccio. They figured out how to marinate and dry meat to change the taste and make the receptors enjoy, like in jamon or marbled beef steaks.

Cooking has grown into an industry with its own canons and rules. For example, Michelin stars have become one of the highest signs of recognition. This hero lived about 100 years ago, then one day he decided to help motorists and created a directory of different establishments where you could eat deliciously. There were gas stations, cafes, eateries, and restaurants.

This is how the famous red reference book and the first stars appeared. They only meant that the prices here were quite high at the start. After 30 years, the story has changed, and Michelin stars began to be assigned to those establishments that especially distinguished themselves, definitely worth visiting at least once in your life. There was a three-star gradation.

One star says that the restaurant is good in its field. Two - that the kitchen is worth attention, and three - that the work of a local chef is worth a separate visit to the institution. How institutions get Michelin stars - no one knows. Because this is like the mystery behind the seven seals, otherwise, all objectivity is excluded. But the fact remains that all restaurants in the world fight for Michelin stars. The most restaurants with stars are in France, but in terms of the number of restaurants with three stars, Tokyo is the leader.

Modern cooking is a combination of technology and creativity

The culinary world is so amazing that now we have both fast-food products and freeze-dried products (soups and instant noodles). Food for astronauts is designed to be eaten comfortably in zero gravity. There are usual semi-finished products that everyone probably has in the refrigerator, and various technologies that allow meat to be softer (vacuum and sous-vide), and fruits and dairy products to be stored longer (pasteurization and preservation). Whenever a new food-related need is born, the world quickly finds a way to satisfy it.

Brief project description

In prehistoric times, people were not picky about their diet, and there was no such thing as proper or improper nutrition. No one thought that food needed to be cooked and that there may be some successful combinations of products that make dishes incredibly tasty. Yes, and cooking shows are worth it. They combined and turned the well-known "bread and circuses" on its head. So, the analysis of the concept of cooking will be fascinating.

Oral manifestation

First of all, these are recipes that were orally passed from person to person. These describe the dishes and the technologies for their preparation, which were originally discussed. Then they were stored on paper and some even turned these into reality with their own innovations. Ideas for creating cooking shows and their different formats can also be considered oral manifestations in cooking.

Actual manifestation

We talk about what we have now and what was before. These are different dishes created according to recipes, restaurants, and places where food is prepared. There are even specialized kitchen utensils used for cooking. There are entire schools that teach cooking, including the hierarchy of cooks and all the staff who work in the kitchen (directly with food). These are different structures that monitor the food quality and develop technological instructions.

Author's analysis

The exploratory concept of cooking is very interesting, especially its author's analysis. Therefore, it is possible to consider it from the all of the various sides of human needs.

Not only from the point of view of satisfying the basic need to get full and get a boost of energy, but also from an aesthetic point of view, to enjoy what you eat and how it looks.

Moreover, if we talk about the phenomenon of culinary shows, then there is a question of eternal "bread and circuses". Therefore, no matter how you look at it, cooking is a huge field for research.

Author's analysis

So, we need to satisfy several priorities. Eat food and make it beautiful. How can we do that?

- First of all, this applies to the food itself. If now we go to shops and markets or use a delivery service, then in the future, we will be able to simply order boxes with products, weekly, monthly, yearly, etc.

- Food can generally change shape, and all the necessary substances can be obtained in tablets or powders to save time, but this will never happen because too many industries make money from food.

Author's analysis

- As for establishments, there are already places with the most daring concepts, those where you can watch how the staff cooks or even participate yourself.

- These can be establishments that broadcast one idea. For example, you come to the Mephistopheles cafe, and they bring you soup in a crystal skull, and the interior itself is artfully presented. That is, here you don't just come to get enough, but also to look at how it's all played out and feel the atmosphere.

- The format of cooking shows may also change. These may well be shows where everyone can participate via webcast or virtual cooking shows where everyone cooks in a virtual world.

Exploring paradigms

Let's analyze cooking from the point of view of the consumer and the entrepreneur. What do they want and where they may not have anything in common?

Who?	What do they do?
Businessman	He looks for ways in which cooking can be filled with new features and monetized. To arouse the interest of the consumer and motivate them to go to this particular restaurant and buy these particular dishes. Such purchases should occur as often as possible.

	The entrepreneur wants to create a unique product that will be expensive, and everyone will want to buy it, but not everyone will be able to afford it. You can create a particular application that will send people the food they need according to specific parameters.
Consumer	He is looking for a product that can satisfy his need, prepared quickly and nicely. Something that he can eat and not waste time cooking. Or, conversely, something that will last longer. Perhaps he will look for a product that combines these qualities, not irritate him, and suit his style, mood, and character.

Neural stories

Cooking, luck, medicine

If you've gone through the concept of destiny before, you know that we combined cooking and destiny. It turned out to be very interesting and informative. As a result, we concluded that even such complex concepts can still be connected and played out originally. But no spoilers for those who haven't

Neural stories

read it yet. So now we will try to combine the concept of cooking and medicine, the concept of cooking and luck.

Luck	You can develop a whole cooking show where two different parties will try their luck. For example, those who prepare a dish must guess which boxes contain what.
	Or choose some food boxes, and cook from them whatever they can, and the judges - to evaluate all of it.
	The main message will be that a person will receive one or another box thanks to luck. Some will be better, some worse; some will not have enough products or devices at all.
	You can also create a restaurant to choose a dish for good luck and try it. You can get something from a Michelin chef or fast food.
	We already observe the combination of luck, for example, when cooks prepare pufferfish.
	Although they undergo special training and prepare for a very long time, the possibility of a mistake can never be ruled out. Luck plays a role here.
	Or, for example, cookies and chocolates with predictions, where you can get a

Neural stories

	positive or negative one.
The medicine	Cooking and medicine go hand in hand. So, for example, it could be a story about a hospital where food is often tasteless.

But here, you can combine the concept of restaurants with healthy cuisine, delivery, and hospitals or sanatoriums, to remove kitchens from hospitals altogether and switch to delivery.

You can also create a restaurant concept where everything will be in a medical style, but for real. Even the dishes themselves would be created so that it is necessary to mix the ingredients in different proportions and add special solutions.

In general, the principle itself would be the same as mixing drugs in medicine. |

By the negativity reinforcement

There is no need to try hard to enhance the negative in the cooking concept. The story has developed so that there have already been several attempts to treat the cooking concept not very well. For example, in the Middle Ages, cooking as a science and art did not exist.

By the negativity reinforcement

People did not have extra money to invest in the development of cooking. People generally ate poorly and often got sick from it. The attitude towards food was such that people ate just not to die.

The peasants worked hard, and the nobles did not feel the need to eat something delicious. Then even the feasts were not distinguished by a special culinary scope. They prepared what they could and the way they knew. Because people were often poisoned, there was no talk of any culinary experiments. Therefore, cooking remained in the shadows for many years.

Later, when people wanted to eat tasty and beautiful food, the first culinary schools appeared. They began to teach how to cook deliciously, and the attitude towards cooking subsequently also changed.

In the future, we will probably have the same attitude to what we are used to now in the same way.

Those culinary items, which we consider unacceptable now, will become very cool in the future. For example, the combination of incompatible ingredients. This could make for a fascinating story.

By rejecting bullying

The concept of culinary innovation by rejecting bullying can be viewed from many angles. However, the main point is that some eat exclusively plant foods, and those cannot imagine life without meat.

By rejecting bullying

Some calmly buy any products in any market and there are others who love exclusively organic; the group of those who most likely like to eat and cook at home, and those who only eat outside. Here we will consider the latter.

Those who like to eat and cook at home, and their arguments.	Those who love to eat outside, and hate cooking, and their arguments.
When you cook at home, you can control the quality of the food and what you eat.	You don't have to spend a lot of time preparing food.
Cooking at home minimizes the risk of food poisoning.	In restaurants, you can try different cuisines that you cannot cook yourself.
When you cook at home, you can enjoy the process itself and experiment.	No need to wash the dishes and put everything away after you cook.
Cooking at home is more convenient and can be done at any time, and it's cheaper.	You can choose different dishes for the family when everyone has different gastronomic preferences.

By the miracle effect and by collaboration

Cooking and the miracle effect are very closely related. Developing this innovation is enough to remember everything that we loved in childhood. Every holiday is always accompanied by delicious food. It is not for nothing that Christmas dishes have always existed in different world cuisines. For example, even the Thanksgiving turkey is a traditional dish with its own story.

Everything is clear here, but how do we experience a miracle in modern times? Easy! For example, you can create a fair to cook their favorite food from games and fairy tales. Would you like to try Harry Potter butterbeer?

Or visit a real fairy-tale feast, which happened in fairy tales about princesses and princes? If we return to the format of cooking shows, then they can also be combined with fairy tales. For example, choose a format where participants cook food from computer games. Or a show where people try completely unusual food and guess the ingredients. By the way, this is also a collaboration.

For example, you can combine the format of talk shows and cooking. This will not be an ordinary culinary show, but a format where you can ask questions and receive products for the correct answers, from which you can then cook interesting dishes. The winner is the team that cooks deliciously and demonstrates its knowledge in different areas.

An interesting collaboration will combine two concepts: cars and cooking. For example, you can run cars around the city that cook and immediately deliver food to places or create a restaurant on wheels that will travel around cities and countries and popularize different cuisines.

By obsessive development

To apply innovation through obsessive development, we need to choose a concept and look at those who use it better. Choose something that we can "steal like an artist" and use it. Then, bring the existing concept to the point of absurdity. So, how to bring cooking to the point of absurdity? What could it be?

Perhaps in the future, we will completely rethink cooking, and it will surprise us greatly; perhaps everyone will have a set of bags with different freeze-dried food, and we will stop going to the cafe.

Perhaps we will change our attitude to food and consume only the substances necessary for the body to function, in the form of liquids. Or, on the contrary, cooking will develop so much that even at school, children will be taught to cook - and not just scrambled eggs, but all sorts of tartars and ratatouille.

I am sure that you got a lot of ideas in the analysis of this concept. Cooking is versatile. It can be twirled in different ways, viewed from different angles, and combined with many other concepts. In the book, I give only the basics and examples with which you can disagree, openly dispute them, come up with your own innovations and implement them. I'll be happy about it. The beauty of conceptology is that it is flexible, like plasticine. Even when I created the book, it changed several times. In the process, I came up with a new innovation and implemented it. The same thing can happen to you. So good luck and don't forget to tell us about your success later.

§8 PHILOSOPHICAL CONCEPTS

Philosophy has always been the science of Wisdom. It was born when people began to think about inexplicable things. People have always tried to find answers to constantly arising questions by observing the world, the environment, their fellow tribesmen, natural phenomena, weather, celestial bodies, and other objects. "Why is it raining? Where does wind come from? Why does the sun always rise in the east and set in the west? Why is everything in the world arranged in this way and not in another way? At these moments, philosophical concepts arose.

How it all began

The beauty of the human mind is that it is curious and observant. Thanks to these qualities, people were once able to create everything that our contemporaries possess. In ancient times, humanity could not explain simple phenomena, so they created myths and legends, invented different gods, and endowed them with superpowers. But, despite the vast and frightening world, they still did not stop asking questions of morality, what is right to do and what is not. Questions arose about what kind of person can consider themself free or happy, Fate, destiny, justice, good, evil, and other categories that exist only in people's minds. No matter how hard they try to inspire or immortalize them in sculptures of deities, they always remain intangible, ephemeral, and non-objective.

For example, philosophy took pride in place in the ancient world and became a way of life. Laws were deduced, state laws were written, and the most daring experiments were carried out. Philosophy is what has given rise to the most advanced inventions but also to the most tragic mistakes. It is the foundation of all discussions, studies, writings, debates, and new views. Of course, philosophical concepts are used in modern times.

No wonder philosophy is the queen of sciences

Philosophy has always been one of the most important sciences, and most importantly - a special form of world knowledge. It helped develop a system of knowledge about the most general characteristics, ultimate generalizing concepts, the fundamental principles of human existence, and the relationship between man and the world.

Philosophy initially raises those questions that most torment the human mind. Even 400 years before our era, Socrates' maieutics led to one of the main features of the modern conceptologist - the ability to doubt everything and look for the maximum number of arguments in favor of various theories, hypotheses, and paradigms. Those areas of knowledge for which it is possible to develop a clear and workable methodological paradigm are separated from philosophy into scientific disciplines. For example, physics, biology, and psychology were separated from philosophy.

The tasks of philosophy for centuries included the study of the universal laws of the development of the world and society, the study of the very process of cognition and thinking, and the study of moral categories and values. Importantly, moral categories are still the subject of controversy and discussion worldwide.

We still cannot clearly define what justice and destiny are. Why, in one case, euthanasia is salvation, and in another case, it is a crime. Why can the death penalty exist in one society and be excluded in another? Why do we all interpret human freedom differently? We can continue ad infinitum. In this section, I want to show and tell you about some of the most controversial concepts that motivate humanity to think, argue, debate, and forever seek a more precise point of view.

§8.1. Should you blame Fate for misfortune? How did the concept of "fate" come about?

How often do we hear or even involuntarily say when we don't get something we want: "Oh, it's not destiny." Sometimes, we are angry at circumstances or because we can't influence something or control something. Sometimes we simply justify our own inaction or even despair. This term is often endowed with tremendous powers. Fate is almost omnipotent because so much depends on it. But have you ever thought about the fact that in this way, you shift your own responsibility to something ephemeral? With a simple "well, it's not meant to be," you deprive yourself of the chance to fight and achieve what you want. However, this does not apply to everyone. What is Fate, and is it so powerful? Can it be changed, curbed, or does everyone live as destined? Let's analyze it further.

Prehistoric fate

Did prehistoric people know about Fate? It's tough to say for sure. After all, Fate is a philosophical concept, complex and multifaceted. To build it, a reasonable person had to gain the experience of several generations, process it, rethink it. Did primitive people have enough such experience? We have no recorded facts about this, but they certainly became the first starting point from which all philosophical judgments began. Already at the time of the very first beliefs, there was some kind of force that determined everything: whether it would rain, whether the harvest would be good, whether a mammoth would be killed, whether a person would die from an illness, and so on. Well, how to say "existed"? It only existed within our group imagination.

Such mythical forces were invented because they could not explain certain natural processes. They simply lacked experience, observations, and understanding of what works and how. In addition, the ancient tribes had more animal instincts, which means they were ruled by fear and anxiety. So, they invented gods and beings who ruled everything. It was easier to live that way and not worry about anything.

After all, when we determine responsibility for others, it is much easier for us to live. So it began: "so the gods decided," "the plans of the gods cannot be understood by mortals," and other beliefs. These beliefs were instilled and passed down from generation to generation for tens of thousands of years. It turns out that the idea that Fate was in charge of everything began to emerge back then.

Antiquity… The time when philosophy was one of the most popular sciences

The ancient Greeks had an even higher god than Zeus: Zeus-Zen. For example, Aeschylus said that Zeus-Zen is the ruler of Fate. In Greek drama, as in the direction and peculiar reflection of the Eleusinian mysteries, they often talked about the mysterious principle of Fate, which guides everything in the world. For example, in the work of Aeschylus, "Prometheus Chained," there was such a moment: the crucified God asked: "Why are you doing this to me?" A mysterious voice from distant heights, over Olympus and over Zeus himself, which reached Prometheus through Hermes, the messenger, the God of Wisdom, answered: "Because Fate wants it, because this is Fate …"

If you look at mythology, folklore, and fairy tales, it is clear that all peoples somehow had a destiny that was even higher than any personified deity. There is a certain, mysterious, mystical, secret principle of Fate, which was not always embodied in some form.

- In Hebrew Kabbalah, Ain Soph (Ein Soph), "Nothingness." It's the same story again: Ain Soph is at the top of the Crown, and he rules everything. He has an impulse that rushes from the heights to reach our world, and in our world, everything that is manifested takes form when certain forces, energies, and matter "collide" due to it.

- In the myths and legends of ancient India, there is also a creature in the pantheon that is beyond rational, intellectual understanding.

- It's all the same in the pantheons of Ancient America and Ancient China: the highest deity is always mentioned without a name and attributes, representing Fate, which cannot be pleaded and cannot be influenced in any way.

Mythology identifies 5 hypostases of Fate

• Distributor. It has no logic; it's just a matter of luck. For some, Fate has prepared a heavenly life; others face many trials, and it is not clear why this is so. Nevertheless, from people who believe in such a universal distribution, you can often hear: "This is my fate."

• Player. This is about those who believe in luck. The wheel of fortune can spin as you like. A person can find himself in a situation that will dramatically change his Fate. They say about such people: they were in the right place at the right time.

• Director. Classic: life is a game, we are all actors, and we get to know certain people for a reason. It's about the power of interaction and events.

• Lender. A person is given a talent, and how he disposes of this determines his life.

• Referee. It all depends on past incarnations. If a person misbehaved and sinned in a previous life, he will be a loser and suffer misfortune.

In the Middle Ages, the concept of Fate acquired religious meanings. "God's will for everything" is a well-known phrase still used today. At the time of the spread of different religions, crusades, and propaganda of beliefs, God and Fate were connected as closely as possible. What is Fate through the glass of religion? Most often, this is a series of tests that need to be passed and endured to prove our loyalty and devotion to God. The Lord predetermines how someone should live, how many trials to send to the lot of everyone, who to reward, and who to teach a lesson. Is it fair that different people receive different trials from higher powers for the same actions? For example, two people committed a theft - it is sinful. But one was caught and punished, and the other got away with it. This is where the critical type of thinking begins its active role. By studying any religious dogma, critical thinking questions all ambiguous or unproven assumptions or sayings. In response, any doctrine or religion provides answers within its own dogma. Thus, Aristotle deduced the law that two objects fall at the same speed regardless of size.

When it was proved that the object itself affects the rate of fall, then within his own scientific dogma, he corrected the interpretation of this pattern by publishing that "objects fall at the same speed, regardless of their mass. But sometimes, the object can be very happy and thus quickly approaches the place of its rest.

When the question arises as to why the same act of two criminals gets a different result, religion explains this within the framework of its teaching. So, there were reasons for this; God decided so, maybe one of the family members sinned a lot, perhaps the person himself does not commit such an act for the first time, or perhaps this is a test for him to check. Initially, the main idea was that everything in the world was decided for us, and we are simply submissive to our Fate.

The philosopher Spinoza, for example, believed that man is only a speck of dust in the Universe. Therefore, it is senseless to expect that this speck of dust can take responsibility for the development of the course of events on a universal scale. But the most critical minds of mankind have always doubted everything. Actually, doubts about the correctness of religious dogmas gave a new impetus to understanding what Fate is.

People could not agree with the fact that everything is predetermined. After all, when you know that this is so, development is impossible. New views and new trends emerged. Observations of people proved: a person could influence his life. What he received depended on his decision and choice. The world began to treat Fate differently, especially after the Middle Ages, when the church ceased to be secular and separated from the state.

At that time already, the main religious components of Fate were formed:

- **Unknowability.**

A person cannot know what the future holds for him. Why did he come to this world, what is his goal, what exactly should he do here, what functions should he perform?

- **Totality.**

Fate concerns everything, all spheres of life and all people. It activates all the universal mechanisms.

- **Independence.**

People cannot influence Fate. No matter how they behave or what they do, it is impossible to change what was destined.

The third one always caused the most substantial doubts. Humanity did not want to put up with everything that was decided for it. Moreover, there were really people who, with their diligence, work and zeal, broke the established stereotypes regarding Fate.

The contemporary definition of Fate

Often, opinions are divided. Fatalists believe that Fate is something destined from the above and cannot be changed. Esoterics believe that Fate is about previous incarnations and karma. Psychologists insist that predestination is founded only on our beliefs, actions, and characters. "Life creators" are sure that there is no fate and that everything is only in our hands. Well, the most cunning fatalists trying to sit on the two chairs don't deny the existence of Fate, but they do not perceive it as something unchanged.

Now there is another, more familiar, gradation of definitions of destiny:

- **Social definition.**
If you are lucky, you have a better chance of living a good life: you have more opportunities to develop and improve if you come from a wealthy family. If you come from a low-income family, it means you have to go through difficulties to get to where the prosperity starts.

- **Psychological.**
Let's take the previous example. The rich child might grow up into a consumer, not know the value of money and absolutely not be able to earn. He might not want to receive an education, live off of someone else, and spiral downwards. At that same time, a person from a low-income family will strive to raise his status, applying more

effort and motivation. His skills will surpass those of competitors with greater opportunities.

- **Eventual.**

Coincidences always exist, although there is an opinion that they are not random (all fatalists think so). This is like in the famous movie "The Butterfly Effect." Even the most minor random events might influence and even move all of the following: conversation, clash of people in a crosswalk, the flap of an insect's wing.

A short project description

In prehistoric times, destiny didn't have any particular form in the human mind, but it was felt, and it was always the case that members of one tribe constantly returned from hikes alive, and the others – always came with injuries. Observing that all childbirth were in the same conditions, cave people came up with the idea that some deity determines how and where the lives of people of that time are moving.

A short project description

There are many different explanations of why it exists and what place man occupies in this relation. From a wide variety of angles, one can spin what happens to a particular being with regard to Fate.

There are many theories about Fate, many different vectors from which it is studied, from the religious to the pragmatic. But our task now is to dissect precisely the concept of Fate.

Oral manifestation

This is when we psychologically reassure ourselves and convince ourselves or others that something bad or good has happened because it is Fate. "I submitted a request to the Universe," "I was helped by higher forces," are all stories about the verbal concept of Fate. Or, on the contrary, we verbally deny the influence of Fate on our success, and in contrast, we say that only with our own strength, persistence, and ingenuity were we able to achieve it.

Actual manifestation

We see variations to the concept of "fate" in recorded prayers, in worship, in attributes. We see it in the belief that certain things can affect the favorability of the outcome of the intended action. For example, a talisman for business success, a "bewitched" moonstone for negotiations to be successful. Even in a religious moment, people worship a deity and go to him to request something good, for the family not to get sick, for a marriage to be successful, etc.

Author's analysis

Remember that whatever concept we analyze, we always keep the brand pyramid in mind and refer to it from time to time. It is important to understand right away that we use the brand pyramid here not according to the stages of perception of destiny, but according to the stages of how one believes in it, the belief that Fate is not one-sided.

At the very bottom of this pyramid is the person who decides nothing for himself and takes no responsibility for his actions or inactions. He justifies any event with the fact that it is Fate's decision. And at the highest level would be the person who would dare to claim that he decides everything, that all his actions and inactions will lead to something, and he is ready to respond to these consequences. Now let's place our fate hypostases in the pyramid:

First (the lowest) - Distributor
There is no logic. It's just a matter of luck. Some people are destined to live in paradise, while others face trials. It's entirely unclear why this is the case. Yet, one often hears from people who believe in this universal distribution: "This is my fate."

Author's analysis

Second - Judge
Everything depends on past incarnations. If a person behaved badly and sinned in a previous life, he will be a loser and suffer misfortune.

Third - Director
Classic: life is a game, and we are all actors. And we get to know certain people for a reason. It's about the power of interaction and events.

Also refers to the third - Player
This about those who believe in luck. The Wheel of Fortune spins in any direction. A person can find himself in a situation that will change his Fate in a very abrupt way. They are in the right place at the right time.

Fourth - Lender
The fourth is the Lender. A person is given talent of some kind, and his life depends on how he uses it.

Fifth (highest) - King of life
This person knows and firmly believes that everything depends only on him. Therefore, even if some circumstances affect his success, he will find and even create other circumstances contributing to his success.

Vulnerable points of the concept

These are all the things we don't like about the concept, which motivates us to improve it later.

What don't we like about Fate?

1. Using Fate, it is possible to control masses, intimidating them.
2. You can convince anyone that they are a loser without strong arguments.
3. You can manipulate facts to justify all by Fate and its influence.
4. Sometimes it is enough to cause an active fantasy in a person by making them believe in the existence of Fate,
5. If you already believe in Fate, then you might not have the courage to take a courageous action.
6. The person who relies on Fate may live a life in vain because he waits all the time for signs of Fate and does nothing himself
7. A person can give his life in the name of the verbal embodiment of the concept of Fate simply by fanatically believing in it.

Study of paradigms

Let's take different types of people who encounter this concept and interact in various ways to understand how these people see the concept from their own perspectives. We try to justify each side, condemn it, and find a middle ground between them.

Study of paradigms
The one who believes that everything depends on Fate

-       Lives by this principle, constantly looking for signs of Fate. The person constantly proves to everyone that only by tempting Fate you can succeed, believing that everything is in the hands of higher forces and does not feel responsible for what he is doing in full. He says, "Well, it not fate," and simply forgets about it if something does not work out. On the other hand, if something works out, he thanks Fate for what he got. He sees meetings not as an opportunity but as a sign of Fate that someone from above has brought two people together.

The one who relies on their own strength

-       You rarely hear from this person that he is grateful to Fate. He believes that only his diligence, hard work, natural charisma, and skills contributed to getting the result. He does not believe that meetings are not random. He knows that if he needs some person in his environment to succeed, he can find the key to get him. This is a man of courage and knows the price of his success. Often, he can go too far and claim that the goal justifies the means, but only because he is fully prepared to take responsibility for his choices and understands the consequences of his decisions and actions. If we consider Fate in a more modern way, it is the interaction of two or more subjects among themselves to achieve a specific goal, not necessarily directly. For example, for someone to buy cups, someone has to produce them. Or being the owner of the plant definitely impacts waste emissions.

They can both get sick with some crap because of it. But it's not a matter of Fate; it's a matter of the consequences of their

Study of paradigms

activities. Both of them. After all, demand creates supply. So although the example is about two people, it can just as easily be projected onto several millions.

Neural stories

Let's take related concepts and connect them.
Machine. Medicine. Cooking.

Fate and machine. You can combine into a concept, for example, a machine that takes responsibility for your Fate. It doesn't just change your Fate, but it changes it for the better.

Fate and medicine. After all, a person's health destiny depends on the indicators of our body, which we and medicine can influence. Therefore, medical institutions do diagnostics first, so they can predict the Fate of our body and allow us to change it if there is a need to do so or to prevent an unfortunate outcome - a kind of fate diagnosis.

Fate and cooking. It is good to play with Fate but in cooking. For example, a game with a dish. A dish with a risk. When you're eating something that needs to be cooked right, and even though you're confident that the cook is super-professional and there's just no way you could go wrong, there's always the risk that something has gone wrong. Even in full-fledged safety, there remains that element of play that already makes the dish, not just a meal but something much more enjoyable with a spice of risk and fear.

Those accustomed to trusting Fate will indeed exult and say that it was definitely Fate's will if anyone was poisoned.

Neural stories

One such example is "fugu" fish dishes. That's the kind of risky cuisine you get.

By reinforcing negativity

This process has been running for a long time, since the moment when people reinforced the negativity of destiny by starting to die for the sake of what they believe in. They justify it by saying that Fate has destined them someplace after death, and nothing is keeping them here, and it's not about facts or arguments; it's about blind faith in something. They are willing to kill each other for what they believe.

Some women have accepted that they will never be able to learn and master professions on an equal footing with men because their Fate has decided. So far, no one has been

By reinforcing negativity

able to correct this falsehood. After all, to do so, one must "break the myth." No one is willing to do this. After all, the myth of Fate is so hyperbolized, deeply rooted in people's mindsets, that it will take many years before it can be eradicated or even just looked at differently. It would have to be a powerful, even aggressive campaign ranging from education to entertainment, to teach people to doubt, think, analyze facts, etc.

By rejecting bullying

The main players on the stage are those who believe that Fate runs everything and those who believe that only people themselves and their actions can affect anything. Those who are sure that nothing is being decided, those who are confident that they fix everything, and those who seem to think that they decide everything but still send requests to the universe.

These can be religious fanatics and agnostics, those who are convinced that a Higher Power rules the world, and those who have convinced everyone of the uniform correctness of the theory of evolution. It can also be ordinary believers who can do everything themselves and pray when they are in dire straits or can't explain anything. What can we do? **For example**, open classes, debate clubs, and talk-shows at school, where everyone can voice their arguments, and everyone else will conclude. The main thing is not to take one point of view as the only true one, but to question all of them. The more arguments there are, the greater the opportunity for others to see different paradigms of Fate and change its perception.

By the miracle effect and by collaboration

This innovation is not possible without human action, and this is the paradox of Fate through this innovation. Because if we do something and people had no influence over it, we just confirm that nothing depends on them, and everything is predetermined by Fate. After all, it was Fate that determined everything so that a miracle happened.

We can take advantage of the fact that the person does something to achieve results, and we can show it in such a light that he did the first, second and third, and he got the result. What is the miracle? The miracle is just the fact that a person overcame his Fate, took some action, and got results.

Without much luck or any connections, he just did it. This story is often used by various coaches and trainers, speakers, and those who help people achieve their goals. They simply show the example of another ordinary person who achieved a result through their own efforts.If this story is also connected and combined with local authorities, it is a great example of collaboration. The authorities will publicize such cases so that as many people as possible will understand that they can influence their own Fate.

They could even incorporate this into the learning process, for example, through various workshops for high school students, and position it so that the earlier you take control of your own Fate, the faster and the more miraculous the results will be.

By reputation enhancement

The basic thesis here is that you can't completely disregard Fate. There is an opinion that it is not Fate that governs everything, but there are orders and secret clubs of those who decide who will be, how many people, and where they will work. Such secret rulers of the world may have technologies and techniques that ordinary people do not even know about. They use them to manipulate people's minds, make people buy specific products at specific times, earn money to spend on particular needs. These people can artificially create anything from hunger to plenty, and they stand at the helm and direct ordinary citizens, like puppets.

But let's leave the conspiracy theories behind. In this vein, Fate helps people not to believe such stories fully. Everyone is equal before Fate, and if such a hidden power does exist, then just as Fate can negatively affect these rulers: they may fall seriously ill, lose their status or influence from independent causes, they may drop out and others may take their place because that is his Fate.

By destruction

What Fate, if there is already a ready-made scenario and everything happens exactly as described in this book. Or it may not be a book, but a computer into which you enter your data, and it shows you what will happen next.

That's the way it's been all of your life. If you drink juice in the morning, it means you'll take the bus to work. If you drink coffee in the morning, it means you'll go to work in a friend's car. There will be a huge number of such chains. The computer will just give you the result, immediately analyzing your choices. And that's all, no more surprises and no more fate. Only the understanding that you got exactly what you chose yourself.

What can we conclude?

The most important thing to take away from analyzing the fate concept is that we cannot rely on one thing alone. There are many beautiful things in the world that our brain can understand and analyze. But at the same time, there are so many unexplored things in the world that we probably can't even think about. But it is conceptologists and their natural tendency to doubt everything that will help us make new stunning discoveries in different fields in the future, even those that seem so far incomprehensible.

§9 ECONOMIC CONCEPTS

The economy is now one of the areas through which the world exists in its present form. However, it took centuries, not just decades, for the world economic order to shape various global concepts. Any financial idea, investment project, or invention is based on fundamental economic concepts, which are the origins of the world economy.

To begin with, let us understand the essence of the economic concept. Its main task is to satisfy humanity's needs, demands, and necessities related to financial phenomena. For example, when we go to the store, we take paper money or a bank card because we cannot buy what we want without it. However, we don't think about how they came into existence, how the world got to the point where today you can buy any product by attaching your phone or watch to a terminal. For thousands of years, we've basically had some kind of coins and paper that allow us to buy anything. But why are there coins or paper?

And if we analyze the store itself? How is it that we come to the marketplace and it has that particular look, with particular products and vendors? If you look deeper, even an ordinary supermarket near your home or even a stall with vegetables and spices at the market - all this is a coordinated work of economic concepts. They have evolved and continue to evolve.

Economic concepts, just like research or philosophy concepts, can include several industrial ones. That said, the emergence of one economic concept necessarily gives rise to the development of another. For example, if the world did not need to invent money, then securities, financial exchanges, stocks, trading, and even cryptocurrency would not have appeared.

If people had not begun to negotiate the exchange of different goods, there would have been no joint bazaars and fairs of various localities. Then there would have been no modern supermarkets, financial markets, alliances of traders, banks, and various financial institutions.

All the named concepts result from two processes: the need to satisfy some need (like all the other concepts) and the need for global regulation of what came out. What does this mean?

It means that if money existed by itself in the pristine form in which it was invented, countries would still be using things that are easy to counterfeit, quick to break, and in principle, easy to obtain. We still wouldn't buy what is available to people in other countries. We would not be able to travel freely around the world, and we would not be able to get something invented in a country other than the one in which we were born.

After all, even exports and imports are also economic concepts. The reason is simple - these concepts came about because there are certain economic needs: getting goods or establishing trade relations with another country, making a profit for yourself, or attracting the attention of those who can give money for the development of your business. It doesn't matter if you're the head of state or if you're a private business owner. The same principles apply here.

Economic concepts and manipulations

Any economic concept will often be closely associated with manipulation or speculation. For example, the existence of the World Bank is itself a phenomenon that is almost impossible to make honest and transparent because a group of people makes the decisions anyway. And does not the same set of principles apply to giving and receiving credit? Even though in today's world, most of these processes rely on computer algorithms, and for many, those processes are unfair and unobjective. Who invents credit rates? Why are they the way they are? What accounts for the terms of interest payments? Why do the reserves of the world bank belong to the private sector? Why is the national reserve of one country or another directly dependent on a group of wealthy families?

After reading this book, the conceptual thinking you develop will help you understand all these questions. You'll be able to see cause and effect relationships, analyze the news and all sorts of information fields that are broadcasting to you, and so on. Even in everyday life, you will be able to use this information and become more aware of how huge corporations can manipulate your mind using the most current economic concepts. In what follows, we will look at specific examples so that you can get the whole picture and figure out almost any economic concept on your own.

§9.1 "Let newspapers be sold and shoes shined!" So where have the shoe cleaners and newspaper sellers gone?

Shoes are an essential element of the wardrobe, without which it isn't easy to imagine modern life. If you look at the images of people of the Stone Age, even they had some kind of shoes. Although in the hot season, they easily went barefoot, walking on their feet on stones and small grains of sand until calluses were formed.

But with the onset of cold weather, people could no longer do without shoes. That is why feet were first wrapped in thick animal skins and tied with self-made shoelaces. E. Trinacus - a historian from Washington University in St. Louis, argues that the first prototype of modern shoes appeared from six to thirty thousand years ago in Eurasia.

Later, the shape of the people's feet of those times began to deform. The scientist concluded that this could happen due to the tight shoes worn at that time. People figured out that it is convenient, comfortable, and much safer when their feet wear something. From that moment on, footwear got a major boost for development. Along with the fact that they were changing, their owners started having new needs. One of these was shoe cleaning, which later grew into a whole phenomenon and gave many people jobs.

"Hey boy, won't you make my shoes shine?"

Although people did not walk barefoot for many thousands of years, there was a mass demand for those who cleaned shoes in the 18th century. Men mostly used the services of cleaners. No self-respecting gentleman could go to a meeting or ball if his shoes were not shining—all the male population and guys from high society needed to visit the shoe professional. Interestingly, in the very beginning, this profession was associated with child labor. Young guys, to help their families and make extra money for their needs, offered shoe cleaning to passersby, often to the wealthy and prominent gentlemen, who were in a hurry on business.

Many street children were not ashamed of this work to make money because that was how they could earn an honest living, and they needed the money just as much as anyone else. Although the pay was not much and the work was not the easiest, a child could provide this service to about 100 people in a day.

What did shoe cleaning look like?

- As a rule, the person was seated in a unique chair, but sometimes the procedure was conducted standing up. The foot was put on a special plank and fixed. Then a special brush was used to remove dust and street dirt, and cardboards were put around the foot in the shoe. This was done so as not to stain the socks. Cream, gutaline, wax were put on the shoes and rubbed in. After a few minutes, they were polished to a shine. The finishing touches were applied with velveteen.

After this ritual, the person went home satisfied with shiny shoes. More often, it was possible to meet shoemakers near buildings, which many people visited, or in places where it was essential to be "dressed up": in squares, markets, and near government offices. People were so fond of shoe-shine men that many turned going to them into a tradition.

Connoisseurs of everything

Shoe cleaners were knowledgeable about everything. After all, their customers were all kinds of people who wanted to talk while the process was going on. So the bright guys gathered information and absorbed the news like a sponge.

Even a legend says that a certain multimillionaire dumped shares of outsider companies just a couple of hours before another stock market crisis. He was immediately asked how he was able to guess what assets to get rid of, and he blamed it on a street kid: "He recently told me he bought shares in railroad companies. Can you imagine a shoe cleaner! That's when I realized that if guys like that were going to the stock market, it was time to withdraw capital.

What did clients do while their shoes were cleaned?

Probably they talked to the cleaner or read. What did they read? Newspapers, of course. Especially when they could hear the boys' voices ringing out the news headlines from every corner, selling newspapers on the street was the second popular occupation young boys used to make money.

Although many of them couldn't even read, they knew very well what the newspaper was about. Newspapers were actively sold in the heyday of printed periodicals (around the middle of the 17th century). But then, of course, they could only be bought at a kiosk or store. It was much more convenient when they could pick up a newspaper just on the way, without going anywhere for that and wasting time.

Selling newspapers was cool

It is believed that the first newspaperman was a boy of ten. His name was Barney Flaherty, and he was hired in 1833. The boy simply responded to an advertisement that said, "Unemployed - staid people can find work by distributing this newspaper. The boy would go to the newspaper stand every morning, take a certain number of newspapers, and sell them to passersby. Barney was the first boy to be hired as a newspaperman. It was official, legal, and paid like any other job. Later, selling newspapers gained incredible popularity worldwide, especially in the United Kingdom, the United States, Canada, Australia, New Zealand, Ireland, and even Japan.

This job was the first paid and affordable work for teenage boys. That way, they could have their pocket money and do useful work instead of hanging around on the streets. Sometimes girls were also newspaper sellers, but it was more often strictly an occupation for boys.

Interestingly, newspapers were not only sold in the streets and squares. In addition to loudly announcing headlines and enticing people to buy the latest press, one of the newspaper vendor's jobs was to distribute newspapers along a certain route to homes and offices. On foot or using a bicycle was a part-time job for teenagers before or after school.

In today's world, both these professions are almost extinct.

The manual labor of shoe cleaners has been replaced by shoe cleaners and sponges soaked with a particular composition. However, these guys can still be found in Asia and Latin America, Afghanistan, India, and Albania.

You can see newspaper sellers in almost any city, for example, near subway stations, and if not them, then the automatic boxes, whereby throwing a few coins into a slot, you can get a newspaper. This is more a tribute to tradition than a necessity because you can easily access any information with the advent of the Internet and gadgets.

Oral manifestation

This helps us quickly get the information we need, sort it out and highlight what we need right now. A person needs to convey important information and emotions about the current course of events, and it also often needs to be done for someone who is far away. Word of mouth is also an example of verbal manifestation.

As for shoe shining, the verbal manifestation is to get clean shoes and not to spend much time on that. There may also be a decorative oral manifestation (clean shoes are more beautiful than dirty ones). Oral manifestation is found in the dress code and the protocol requirements oblige a person to wear shoes shined to a shine.

Actual manifestation

We talk about what we have now and had before. These are the specially trained carrier pigeons who brought news. These

Actual manifestation

messengers who delivered scrolls to their lords, mailers, newspaper subscriptions, or messengers with news feeds.

Particular sites publish only news. It's shoeshine machines, various tools to protect against dirt, even special shoe covers, which are put on when it's bad weather in the city. There can even be moisture sprays or dust creams.

Author's analysis

These both concepts existed at the same time. They were complementary and independent. It is difficult to make an authoritative analysis for concepts that have already outlived themselves, but we will try it. The need for information and clean shoes went nowhere. **So how do people satisfy it now?**

- Some special firms and companies are engaged in chemical shoe cleaning. They position themselves as such that they can clean anything literally, even if your favorite shoes have had dirt on them since 2007.

Author's analysis

• The firms have developed special solutions that repel moisture and dirt. This way, they have eliminated the need to clean shoes at all.

• In many hotels, even now, there are machines, where you just stick your shoe with your foot, and the mechanism with brushes comes into action by pressing a button and cleaning your shoes.

What else could there be? Perhaps there will be no need to clean shoes at all in the future. There will be nanobots that you can "put" in your sneakers, and they will clean them in 2 seconds.

• As for newspapers, there are many different options. From news channels in your favorite messenger to digital variations of glossy magazines.

• The idea we see in the Harry Potter movies is also an interpretation of how information is presented and communicated in the modern world. After all, to think that we really take tablets in our hands and read the news with pictures that come to life, only we call them videos or gif-animations.

Ways of concept development

So how can concepts evolve that have practically outlived themselves and remain only in the format of good old traditions or some kind of customized stories?

If we're talking about conveying the current state of

Ways of concept development

information - games, music, graphics - you can go far beyond the usual messengers. For example, one can imagine the existence of such virtual offices, where everything is created in augmented reality. This is for those who may miss the office routine but are not ready to spend time on the road.

What does it look like? You put on augmented reality glasses and enter a world where everything is set up like in the office. You can even go from office to office and communicate live. Create your own avatar image, which will be alive in the virtual world. Perform the same actions as you, adopt your habits...

This format can be more than just a working format. It can be turned into an area of relaxing and communicating. Imagine that to talk to a friend who is on the opposite side of the world, just go to the app, connect your avatar, and that's it. Sing songs around the campfire, talk, look at the stars with the whole party. And pass the information on to each other.

If we talk about ways of developing shoe-cleaning services, we can only assume that we will stop using shoes in the way they are now. Accordingly, we will need to clean in a completely different way. For example, it could be a self-cleaning film that can simply be sprinkled on the shoes.

Study of paradigms

Let's take different types of people who encounter this concept, interact in different ways, and understand how these people look at the concept from their perspectives. We try to justify each side, condemn each side, and find a middle

Study of paradigms

ground. **In this case, we can only talk about the concept of transmitting the information.**

The one who accepts information	• Wants information to come quickly • Wants to find the suitable topics quickly • Wants to get reliable information • Wants to get information no matter where you are • Wants the information not to be spammed and go where they need it. • Does not want to wait for large amounts of information to be downloaded from the sender
The one who sends information	• Doesn't want to worry that the recipient won't get what was sent to him • Does not want to worry that important

Study of paradigms

	information may be intercepted • Afraid the information will lose relevance before it gets there • Afraid of making a mistake and the info will go to the wrong recipient

Vulnerable points of the concept

What don't we like about shoe shining and transmitting information?

1. That information may be unreliable.
2. You have to wait a long time to get the right information.
3. It takes a long time to get what you want and try different queries.
4. Sometimes you only remember some passages or a motif (like a song), but that's not enough to find what you want.
5. You have to use complicated encryption keys to send secret messages.
6. It takes time and physical effort to clean your shoes.
7. It is not always possible to clean all the dirt with one product.
8. You need to spend time and money on special dry-cleaning shoes, which is not always convenient.

Vulnerable points of the concept

9. Some dirt cannot be cleaned at all. You either have to buy new shoes or completely restore the old ones.

Neural stories

Let's take related concepts and connect them:
Tree, Crowd, Newspapers, Shoe Cleaning

Wood-Newspaper-Shoe cleaning
- You can think of a particular service of such trees-postmates that deliver parcels. A tree with a screen that runs on natural energy. At any time, you can find out on such a screen by simply requesting a screen, simply by submitting a request.

- The very idea of a forest with trees as a prototype of a certain network that transmits information. It can be divided and conventionally depicted on several levels. For example, branches are one system of information, with small messages. The leaves are dotted with short messages. The tree's trunk is a large array of data connected, and the roots are the system that supports and feeds it all.

Neural stories

Crowd-Newspaper-Shoe Cleaning
- The idea of conveying information at scale can be very interesting. We all know about the Secret Santa tradition on New Year's Eve. Imagine how cool it would be to organize an entire worldwide network in this way. A person would first get an invitation, then, for a nominal fee, subscribe, make his wishlist, and then get another member's wishlist. One can also exchange information in this way.

So, you can do for someone else a real miracle and get yourself a nice gift. The main thing in this story is to discuss and agree on the average cost and check in advance so that no one gets hurt.

By reinforcing negativity

Here's where the revelation awaits you. Not necessarily all innovations have to be inherent in every concept. You don't have to rack your brains about applying this or that innovation. After all, sometimes the answer lies on the surface, and it's very simple.

In the case of outdated concepts, it is often the case that there is nothing to amplify. Since the concept has not been used in its original form for a long time, we do not strengthen the negative here because there must be a real story that we can then use. There's nothing to amplify since there can't be one, which is a normal story.

By rejecting bullying

We are looking at two audiences who try to bully each other. They are literally fiercely defending each other's points of view, which are opposed. So, we could be talking about those who generally loathe the use of artificial media and prefer to communicate and share information live, and those who hate live conversations and prefer to pass on info via text.

Those who love to communicate in person and their arguments	Those who like to share by text and their arguments
Live, you can see a person's emotions and understand if they lie.	Passing information through media and messengers faster.
Live communication energizes you and helps you better understand a person's emotions.	You don't have to talk to people you dislike and waste energy on.

By rejecting bullying

This minimizes the likelihood that information will get to the wrong recipient by mistake.	You can record what you say and use it as proof that you are right.
You don't have to worry about the safety of information delivery because everything happens directly.	You can store in one place a lot of information in different formats for later (text, photo, video)

By the effect of miracle

Do you remember how we all tried on the roles of fictional characters when we were kids? We played our favorite cartoons, imagined ourselves as Mickey Mouse or The Lion King, and the tradition of dressing up for Halloween and New Year's Eve masquerades. What if we told you that it could be implemented?

Imagine that it would be possible to create a world with a thousand characters and fully reproduce their characteristics for a second. It wouldn't just be a character (as in a game where you play as someone) - it would be a full-fledged personality when you imagined how great it would be to be able to race between skyscrapers like Spider-Man or fly like Superman or become invisible. In this world, you can try and experience everything. In the same way, you will be able to exchange information. You can create special mail for messages, rooms of interest. It may completely overturn the world of human perception and the transfer of information from person to person and not only amuse your inner child but even do something revolutionary that is unimaginable.

By improvement of reputation

We can play on the fact that the other concept has done it before, but we've made it better. For example, if they passed information in letters that were kept in special boxes in the past, then we can make those boxes fly themselves. This means that it may well change and reincarnate into a structure that simply floats in the air in the future.

For example, the information was transmitted orally only in the past, and then came paper and pencils. Then electronic media appeared, which greatly excited the world, and then the transfer of information may well become something that doesn't require separate media.

> **To improve the reputation of your concept, you can use the good reputation of another concept.**

And this is NORMAL. And it's LEGAL.
In the same way, once a phone with buttons turned into a touch phone.

By collaboration

We create a symbiosis with another concept, improving its functionality but not changing anything categorically. For example, we could consider collaborating the concept of transmitting information via messengers and adding a modified format to the messenger itself.

When a lot of information needs to be stored there, you can just create a separate gadget that contains everything you need like a flash drive, but with a controllable screen. Or, for

By collaboration

instance, combine a flash drive with a password generator. Create such a thing that generates passwords and saves them. It could be stored separately or attached as a keychain.

By obsessive development

We see that someone uses the concept better than us, or just like us. We take what we know about the concept, improve it, and develop it comprehensively. It can be a shoe cleaning or a news broadcast no longer like its first form. To begin with, we improve it to the point of absurdity. That is, it's no longer boys cleaning shoes but flying cars that you can call out with a snap of a finger or a special sign.

Or, in general, holographic images of these same boys from antiquity. As for the transmission of news, it is possible to make it so that a person receives information directly into the brain, without a medium.

On this great analysis, we will conclude our book. After reading all the pages, the main thing you should have realized is that conceptology is a flexible science. That in the process, you will have new ideas and discoveries. Be sure to write them down, record them, sketch them out. You may come up with some new format of innovation, and then they will be used worldwide. Perhaps you will discover something in your activity that you have not thought of before, and that will be wonderful.

§10 REALIZING POTENTIAL

Still, not everyone will become a professional conceptologist. That's the opening in the last pages of the book. "Why didn't you say so before? What should I do now, since I've already tested everything, practiced it, gone through the exercises? All for nothing? Are there any other pitfalls I should know about?"

That's probably what you thought. But no, it was not for nothing. Although the statement is true, not everyone who has read the book and even worked with workbooks (you can also find them on our official site or other online stores) will eventually become a full-fledged conceptologist. But absolutely everyone will be able to get out of the book what they will need in the future to achieve their goals. Most importantly, it provides new skills for working with concepts. As it turned out regularly in my experience in this field and in teaching new students, many people at the beginning of the book did not even know the meaning of the word "concept," or they interpreted it rather ephemerally. And this is normal. After all, in any case, conceptology is useful to everyone, regardless of your field, ambitions, or plans for life.

Even if you're not planning to go deep into conceptual studies, you can still make it your main, because you've still gained an irreplaceable experience that you probably couldn't find anywhere else. If you didn't mindlessly turn the pages but rather did exercises and tried to comprehend or even discuss our analysis of concepts, you gained skills and knowledge that have changed you both as a person and as a professional.

Conceptualization is a story about cross-collaboration

The skills you've learned can already be applied to your professional life. For example, if you work with finances, you will now optimize your work by decomposing it into concepts.

After working through them using my methodology, you will find ways to solve those problems that previously seemed difficult and unsolvable.

Maybe you will invent a new program that will simplify the calculations, maybe you can give developers in your company an idea so that they can write such software which will help you to spend 10 minutes instead of 2 hours on routine calculations. Perhaps you will look at your company processes from a new angle and see all the jumps and dips you didn't notice before. In general, no matter what area of your work is concerned - conceptology is applicable everywhere! After all, you've traced your own path through the entire book.

Together you and I have already taken apart several concepts from different fields. So, what prevents you from continuing to do the same thing in your field? Whatever you're facing, whatever the task at hand, you now have this powerful tool to help you solve it. Getting ahead of yourself, you are now holding in your hands what will become your reference point for any future activity. Even if in 10 years you want to radically change something in your life (for example, working now as an accountant, and after a while will go into the design), then you will do it much easier than those who do not have in their hands the working arsenal of the conceptologist. You'll be able to raise your status as a professional if you know how to solve a problem in more than one way.

You'll be able to offer alternatives to your managers or business owners depending on their needs and capabilities. You will be able to find solutions where others will fall into a dead-end stupor. You will be able to share your skills and train others in your field. You will be able to become a mentor and an authority for those who are just beginning to develop themselves in your field and in their understanding of basic conceptualization.

§10.1 Conceptology as a profession

How will conceptology develop further? Of course, the book is not everything. It's just the beginning. We also have a website dedicated to the support and development of conceptology. There you can find our workshop book with an elaboration of different individual concepts (so you can concentrate on the area that you need and find useful), courses on different levels of professional development for conceptologists, and scientific publications. Also, we already have an Instagram page where we regularly share interesting concepts and little-known facts about common concepts at the time of writing. I plan to launch a YouTube channel where I will share free additional information from this field with you.

I have repeatedly said and will continue to say many times that everything you have learned from this book (and further learn from my other resources) - you can safely distribute, use in your activities, send links to useful data, self-train others, apply in practice in their businesses, use as a tool in the current activities, save and distribute, record, forward, etc., but only under one condition - be sure to cite the original source. This prevents piracy of intellectual property rights and helps us maintain the integrity and authenticity of the information we distribute. This way will affect the healthy development of the conceptual community and actively increase our ranks.

I'm confident that in the future we will organize different kinds of conferences, meetings and share not only our experience but also our achievements. I'm sure there will be a lot of them. Maybe, as you are reading this book, we have already created a calendar of events and regularly hold seminars and conferences worldwide. Who knows, maybe right now someone is holding the book in his hands who will be inspired and will make a colossal discovery that will change the world we are used to. For example, to figure out how to travel between galaxies faster or how to clean up the world's abundance of waste, and for such a person, I will say that it is not worth dwelling on this book or even going through the workbooks.

Moreover, you shouldn't even dwell on the seminars or those courses that I prepare with the book. If you feel the fire of conceptology burning inside you, read related literature, develop your knowledge of languages and world history, learn to play different musical instruments, learn to draw and sculpt. Do not stop with the received knowledge and instead burn with the desire to learn and try more and more.

To those who chose the path of a professional conceptologist, I can say that your journey does not end on the eternal improvement of skills and providing these services to various businesses! How about your statuses on social networks, where you will certainly indicate that you are a conceptologist? To begin with, you must go the way of misunderstanding and perhaps even rejection. After all, this is not yet taught in universities. The very first ones will be the hardest. But you already have ready resources, a book, materials and articles, social networks, and an existing community. It will be easier for you than it was for me, and it will be easier for the next one than it was for you. And it will be easier THANKS TO YOU! Every piece of information has an addictive phase, and it will take some time for humanity to get used to the existence of this scientific field and then recognize it. But the sooner you join this journey, the greater mark you will be able to leave.

In every science, you have a high probability of coming up with new solutions. And in our case, it's innovative solutions for new innovative solutions. As strange as it sounds, it is inspiring. You are no longer just a user of our knowledge and skill - you are a concept discoverer and creator of the future tools.

You can describe your experiences in articles and send them to us for publication on our portal. You can also publish your experiences on other portals in the field of conceptology. You may write not only articles but books as well. As you explore new conceptual fields, you can spread your unique conclusions and thus make our community more multifaceted and as active as possible.

Perhaps you lack like-minded people. Then you can find a club of conceptologists in your country or city. If one does not yet exist, you can start one. Contact us through official sources, and we will definitely help in its creation and development. You can use our open materials for translation into other languages, distribute them through new pages in social networks, your own websites, YouTube channels, or in any other way you wish. I'm all for distribution. That is my mission in this, so far, new scientific direction. Take all the available free materials and distribute them by any means, using any channel, making them more transparent and more accessible to all humanity. The most important thing is to respect the copyright of primary sources.

In conclusion, don't be afraid to experiment and question everything around you. In search of truth, find it, but doubt it and question it. Look for new facts that others believe and try to understand them too. **Look at familiar things through the lens of the unfamiliar.** Do my exercises regularly and share them with your friends or students. Also, share your progress on social media and other resources. Don't be afraid to **start or suggest something new, even if it seems odd.** That's what the most explosive discoveries come out of! In addition, **as much as I believe in conceptology** and its potential, **as much I believe in everyone** who is fired up about this book and science field.

So, you just have to know that you are not alone!

USED SOURCES AND LITERATURE

http://crydee.sai.msu.ru/ak4/Chapt_2_35.

https://lexicography.online/etymology/а/астра

http://old.ihst.ru/aspirans/astronomyia.htm#_Toc100630698

https://sites.google.com/site/astronomia2410/home/our-story-1

http://www.inggu.ru/upload/lectures/лекции%20по%20астрономии%202020%20-%20физика.pdf

http://mmf.pskgu.ru/ebooks/astros/9608_G.pdf

https://starcatalog.ru/osnovyi-astronomii/vidyi-astronomii-i-ih-podrobnoe-opisanie

http://window.edu.ru/catalog/pdf2txt/952/20952/4165

https://classes.ru/all-russian/dictionary-russian-foreign2-term-4659.htm

https://www.efremova.info/word/klub.html

http://ruka-na-pulse.ru/news/detail.php?ID=709

http://noskol-crb.belzdrav.ru/psikhologicheskaya-pomoshch/stress-i-distress-.php

https://medcentr.zp.ua/what-is-stress/

http://mopnd.ru/index.php/depressiya/stress

https://www.smclinic-spb.ru/sm-info/1628-stress-po-polochkam

https://openknowledge.worldbank.org/bitstream/handle/10986/13807/55403Russian.pdf?sequence=3&isAllowed=y

https://militaryarms.ru/vedomstva-i-organy-upravleniya/vsemirnyj-bank

https://www.vsemirnyjbank.org/ru/about/history

https://osvita.ua/vnz/reports/bank/19823

https://www.hneu.edu.ua/shho-take-svitovyj-bank

https://anews.com/novosti/121835070-vsemirnyj-bank-chto-jeto-takoe-gruppa-vsemirnogo-banka-i-rossija.html

https://militaryarms.ru/vedomstva-i-organy-upravleniya/vsemirnyj-bank/#h2_2

http://www.sovslov.ru/tolk/gazon.html

https://educalingo.com/ru/dic-en/grass

http://gazoni.com.ua/pages/view/istoriya_gazona

https://samaragazon.ru/istoriya-gazonov

https://igazon.ru/article/zachem-nuzhen-gazon

http://www.seeds.ru/art-38-interesnaya-statya-pro-gazony-istoriya-poyavleniya-htm

https://dictionary.cambridge.org/ru

https://samaragazon.ru/upload/iblock/ros/Roman_o_rose.pdf

https://lexicography.online/etymology/

http://samdizajner.ru/obzor-naibolee-populyarnyx-vidov-gazona-ustrojstvo-gazonov-i-foto.html

https://mirgazon.ru/blog/istoriya-poyavleniya-gazonov

http://www.seeds.ru/art-38-interesnaya-statya-pro-gazony-istoriya-poyavleniya-htm

https://classes.ru/all-russian/russian-dictionary-Vasmer-term-2342.htm

http://ztchess.inf.ua/?p=4262

https://historygames.ru/nastolnyie-igryi/istoriya-go.html

http://www.gambiter.ru/go/item/159-history.html

https://sites.google.com/site/4kursmath/igry-i-matematika/istoria-igry-go

https://www.sente.ru/ob-igre-go/

http://wmsg.ru/go/go-equipment/

https://clubgo.ru/go-and-ai/

https://burunen.ru/news/society/58807-go-samaya-uvlekatelnaya-igra-v-mire/

https://kuking.net/10h.htm

https://receptino.ru/141-istoriya-kulinarii

https://dom-eda.com/Lyalya/2013/12/13/istoriya-vozniknoveniya-i-razvitiya-kulinarii.html

https://cookzametki.com/

https://makeitshow.com.ua/ru/news/kto-pridumal-zvezdi-mishlen-i-za-chto-ih-na-samom-dele-dayut
http://priprava.by/istorija-pojavlenija-prjanostej-i-specij/

https://biletsofit.ru/blog/kulinariya-kak-iskusstvo-bitva-za-shedevr

http://vovet.ru/q/proishozhdenie-slova-kulinariya-kak-ono-popalo-v-russkij-yazyk-2y3.html

https://snob.ru/style/istoriya-naruchnyh-chasov/

https://juvelirum.ru/vidy-juvelirnyh-izdelij/yuvelirnye-izdeliya-chasy/istoriya-naruchnyh-chasov

https://montre.com.ua/novosti/istoriia-naruchnyh-chasov

https://www.breguet.com/ru/

http://www.sekunda22.ru/informatsiya-o-chasakh/131820/

https://chasik.com.ua/news/kak-poyavilis-pervye-naruchnye-chasy-istoriya-vozniknoveniya-hranitelej-vremeni/

https://elemian.medium.com

https://316.watch/blog/istoriya-naruchnykh-chasov/

https://www.optix.su/blog/istorija-solncezashhitnyh-ochkov/

https://refaced.ru/blog/pervye-solntsezashchitnye-ochki.html

https://happylook.ru/blog/solntsezashchitnye-ochki/istoriya-solntsezashchitnykh-ochkov/

https://zaidiuvidish.ru/o-poleznom-i-krasivom/istoriya-solntsezaschitnyih-ochkov

https://happylook.ru/blog/solntsezashchitnye-ochki/istoriya-solntsezashchitnykh-ochkov/

https://www.marieclaire.ru/moda/ot-gladiatorov-k-aviatoram-istoriya-solntsezaschitnyih-ochkov/

https://lexicography.online/etymology/д/деньги

http://tolkslovar.ru/d2104.html

https://kartaslov.ru/значение-слова/деньги

https://arzamas.academy/courses/74

https://www.popmech.ru/technologies/6419-tverdaya-valyuta-dengi-dengi-dengi/

https://fortrader.org/eto-interesno/kak-poyavilis-dengi.html

https://investingnotes.trade/kak-poyavilis-dengi.html?__cf_chl_jschl_tk__=30a1266f85bbe087bfdeac59a5f8e812db163ea0-1613655973-0-AYBi4w8IReJqfXHdKzsV02J6VNVmS1VsFJ_N0W8AIf8XIOPgdl2aW03MudVY8kvupJ1F_ycQHkiTP2B3-v3A5mZ0C6X1HKo_ykPo2hwFuy8Pi_4TnN6RDqm_k0OVNjHPHOx5qSDBkKL78BT3dm7_lYeP2frztnoyLdNvLmebOr3rp4AWvyAk5sGlPFP3Udy7R6gNGAJD8D9qT_GTXcFiX7UYEf8zsog9nsE9mjrzuv_c9twLFWYE-t-z2T2KHK_ucbVN7pvbgH9-90DMUg7-07npLXlWSDj9MzYFFZdjcAdObhjK97l-ZeKm9oIC3a-E4C-8YIoE1AdZqjO8PbukAuI

https://fostylen.com/archive/gde-i-kogda-pojavilis-pervye-dengi/

https://www.monetnik.ru/obuchenie/numizmatika/istoriya-deneg/

https://vtbrussia.ru/tech/tri-pistolya-pyat-eskudo/

https://infoekonomika.ru/ehkonomicheskie-discipliny/investicii/fondovye-birzhi-istoriya-vozniknoveniya-status-razvitie/

https://articlekz.com/article/4578

http://www.aup.ru/books/m225/5_1.htm

https://nettrader.ru/article/slovar/

https://classes.ru/all-russian/russian-dictionary-Vasmer-term-928.htm

https://ecanet.ru/word/%D0%91%D0%B8%D1%80%D0%B6%D0%B0

https://portal.tpu.ru/SHARED/m/MIKITINA/Teaching_materials/Tab1/Stock_exchange.pdf

https://kakrasti.ru/investisii/dlya-novichka/istoriya-vozniknoveniya-fondovogo-rynka-kak-poyavilis-pervye-birzhi/

https://ffin.ua/ru/blog/articles/investopediia/post/istoriia-birzhovoi-torhivli

http://www.bibliotekar.ru/finance-3/86.htm

http://

ymadrodd.blogspot.com/
2015/08/kredit.html

https://ecanet.ru/word/

https://classes.ru/all-
russian/russian-dictionary-
Vasmer-term-6100.htm

https://
www.kommersant.ru/doc/
2883972

http://tristar.com.ua/2/
rdoc/
istoriia_kreditovaniia.html

https://globalcredit.ua/
novosti/kto-pridumal-kredity-
istoriya-kreditovaniya-ot-
drevnih-vremen-do-20-veka

https://discovered.com.ua/
glossary/istoriya-kredita/
https://credits.su/
magazine/others/istoriya-
kredita-prichinyi-po-
kotoryim-voznik-kredit/

https://www.banki.ru/
wikibank/istoriya_kredita/

https://globalcredit.ua/
novosti/kto-pridumal-kredity-
istoriya-kreditovaniya-ot-
drevnih-vremen-do-20-veka

https://mebel-news.pro/
articles/the-history-of-
furniture/table-history-
ancient-tables-and-their-
modern-counterparts/

https://faqed.ru/history-
historical-notes/stul-istoriia-
proishozhdeniia.htm

https://sites.google.com/
site/muzejistoriizilisa/
stolovaa/istoria-
vozniknovenia-stulev

http://
www.tsuricom.com.ua/publ/
5-1-0-108

http://ec-dejavu.ru/s/
Stol.html

https://mebel-news.pro/
articles/the-history-of-
furniture/table-history-
ancient-tables-and-their-
modern-counterparts/

https://lexicography.online/
etymology

https://www.znajkino.ru/
english_rus_chair.htm

https://tvorcheskie-
proekty.ru/node/695

https://fastpost.org/at/blog/
istoriya_vozniknoveniya_stula
_v_ochen_kratkom_izlojenii

https://faqed.ru/history-
historical-notes/stul-istoriia-
proishozhdeniia.htm

https://billionnews.ru/3969-
istoriya-vozniknoveniya-
stulev-11-foto.html

https://kartaslov.ru/значе-
ние-слова/каблук

http://что-означает.рф/
каблук
https://
lexicography.online/
etymology/к/каблук

http://cocktail-shoes.ru/?
show_aux_page=35

https://www.spletnik.ru/
blogs/govoryat_chto/
142853_kak-poyavilsya-kabluk
https://itaita.ru/news/
istoriya_vozniknoveniya_kabl
uka

https://www.vokrugsveta.ru/
quiz/566

https://itaita.ru/news/
istoriya vozniknoveniya kabl
uka

http://nakonu.com/
2015/09/21738

https://respect-shoes.ru/
articles/kto-pridumal-
kabluki/

https://www.vogue.ru/
fashion/news/
pochemu_zhenschiny_tak_pol
yubili_tufli_na_shpilke

http://www.dancedance.ru/
soveti-novichkam/
tancevalnaya-obuv.html

http://www.danceduet.ru/
index.php/poleznye-sovety/
vidy-kablukov

https://danceshop.ru/
poleznaya-informatciya/
vysota-i-forma-kabluka-dlya-
tantsev

https://www.marieclaire.ru/
moda/obuvnoy-slovar-10-
vidov-prekrasnyih-kablukov/

https://burdastyle.ru/stati/7-
vidov-kablukov-o-kotoryh-
polezno-znat-/

https://oknasmart.ru/
slovar_terminov/okno-
etimologiya-slova

https://www.stroypraym.ru/
2011-07-04-13-26-35/okna-
dveri/2040-kak-poyavilis-
okna-v-domah.html

https://hatka.org/articles/
pridumanyi-plastikovyie-
okna/

https://www.yrossi.ru/
istoriya-sozdaniya-okon.html

https://lexicography.online/
etymology/krylov

https://ornet.com.ua/
articles/okonnaya-
istoriya.html

http://www.vashdom.ru/
articles/fenstr_3.htm

https://oknakr.dp.ua/
stati/kogda-poiavilis-stekla-v-
oknah

https://mirnovogo.ru/okno/

https://lexicography.online/etymology/vasmer/л/ложка

https://villagrazia.ua/blog/cutlery-history

https://autogear.ru/article/276/829/vilka---eto-znacheniya-i-proishojdenie-slova-vilka-primeryi-upotrebleniya/

https://foodandmood.com.ua/rid/news/709423-vilki-da-lozhki-kak-poyavilis-stolovye-pribory
https://vazaro.com/forcustomers/articles/720/

https://ekodomus.ru/blog/raznoe/istoriya-stolovykh-priborov/

https://bellezza-storia.livejournal.com/63664.html
https://lifestyle.segodnya.ua/lifestyle/fun/kak-vy-dorogi-k-obedu-istoriya-stolovyh-priborov-652465.html

http://www.koryazhma.ru/usefull/know/doc.asp?doc_id=114

https://lexicography.online/etymology/в/вилка

http://www.endic.ru/rusethy/Vilka-4488.html

http://endic.ru/rusethy/Nozh-2322.html
https://classes.ru/all-russian/russian-dictionary-Vasmer-term-8468.htm

https://lexicography.online/etymology/н/нож

https://kedem.ru/various/istoriya-stolovyh-priborov/

https://foodandmood.com.ua/rid/news/709423-vilki-da-lozhki-kak-poyavilis-stolovye-pribory

https://lexicography.online/etymology/
https://www.hisour.com/ru/origin-of-the-umbrella-china-umbrella-museum-48261/
https://hystoryfashion.ru/

http://xn----dtbjalal8asil4g8c.xn--p1ai/galantereya/istoriya-zonta.html

http://endic.ru/enc_fashion

http://endic.ru/enc_ancient

https://librebook.me/the_posthumous_papers_of_the_pickwick_club/vol1/1

http://blog.aquamir.kiev.ua/

https://kartaslov.ru

https://vencon.ua/articles/istoriya-konditsionera-sozdanie-i-razvitie-otrasli

https://tehnikaland.ru/klimaticheskaya-tehnika/istoriya-konditsionerov.html

https://www.amegaklimat.ru/polezno/tekhnologii/kto-i-v-kakom-godu-izobrel-konditsioner/

https://euroclimat.ru/presscenter/articles/88/

https://zoom.cnews.ru/publication/item/2037

http://scsiexplorer.com.ua/index.php/istoria-otkritiy/1830-istorija-konditsionera.html
http://www.berlogos.ru/article/sredneaziatskie-portaly-srednevekovye-kondicionery/

https://otvetus.com/chto-delali-drevnie-civilizacii-chtobi-sohranyat-hladnokrovie-te-ispolzovalis-li-primitivnie-formi-kondicionirovaniya-vozduha-i-prodolzhali-li-oni-ispolzovatsya-v-srednie-veka-epohu-vozrozhdeniya-i-t-d-93652

http://5klass.net/informatika-6-klass/noutbuka/003-Proiskhozhdenie-slova-noutbuk.html

https://kartaslov.ru

https://ref.ua/articles/istoriya-razvitiya-pk-ot-ogromnogo-shkafa-do-sovremennogo-kompyutera-za-75-let/

https://www.dw.com/ru/как-лень-заставила-немца-изоб-рести-первый-компьютер/a-5732057

https://shalaginov.com/2019/08/16/6268

https://ichip.ru/tekhnologii/istoriya-kompyutera-ot-kalkulyatora-do-kubitov-245122

http://marsiada.ru/357/465/728/notebook/

https://24smi.org/news/26114-kto-i-kogda-izobrel-pervyj-noutbuk-v-mire_facts.html

https://hi-news.ru/banderolka/kratkaya-istoriya-pochty-chast-pervaya-ot-signalnyx-kostrov-do-nashix-dnej.html

http://postlite.ru/history_post.html

http://fmus.ru/article02/Sorkin.html#A12

http://mirmarok.ru/book/gl01.htm

https://icgol.ru/raznoe/lyubopyitnyie-faktyi-iz-istorii-golubinoy-pochtyi.html

https://world-post.org/rus/novost/?n=9

http://post-marka.ru/stati/golubinaya-pochta-odin-iz-sposobov-pochtovoy-svyazi.php

https://agronomu.com/bok/7035-kak-ranshe-rabotala-golubinaya-pochta.html

https://habr.com/ru/company/megafon/blog/192638/

https://ria.ru/20130129/920119858.html

https://www.arthuss.com.ua/books-blog/istoriya-feykovykh-novyn

https://vm.edupressa.ru/gazeta/otkrytyj-urok/fabrika-novostej-gazetnyj-mir-stolits/

https://newstyle-mag.com/history-of-tv/

https://журналистика-обучение.рф/vsya-istoriya-smi/

https://newstyle-mag.com/history-of-press/

https://dic.academic.ru/dic.nsf/ruwiki/293062

https://knife.media/what-is-justice/

https://skepdic.com/wp-content/uploads/2013/05/4263Nozik.anarchy_state_utopia.pdf

https://glosum.ru/

https://gtmarket.ru/library/articles/2663

https://kartaslov.ru/

https://iphlib.ru/library/collection/newphilenc/document/

HASH9f5facde7eb2c0a9b05e9f

https://www.kom-dir.ru/article/2849-kollaboratsiya

http://www.fingramota.org/teoriya-finansov/ustrojstvo-fin-sistemy/item/830-federalnaya-rezervnaya-sistema-ssha-istoriya-razvitiya-tseli-i-zadachi

https://dic.academic.ru/dic.nsf/ruwiki
https://www.economics.kiev.ua/index.php?id=1022&view=article#jef

https://vc.ru/finance/117361-frs-ssha-kak-ustroen-samyy-vliyatelnyy-centrobank-mira

https://www.banki.ru/wikibank/federalnaya_rezervnaya_sistema/

https://berg.com.ua/world/federal-reserve/

https://smart-lab.ru/blog/62636.php

https://goldenfront.ru/articles/view/25-bystryh-faktov-o-federalnom-rezerve-kotorye-vy-dolzhny-znat/

shubki.info/kozhanaya-moda/kozhanaya-obuv/103-istoriya-obuvi-ot-drevnosti-do-nashih-dney.html

https://fishki.net/2406424-kuda-propali-chistilywiki-obuvi.html

https://fishki.net/2406424-kuda-propali-chistilywiki-obuvi.html

https://lisette-paris.livejournal.com/20862.html

https://fortrader.org/eto-interesno/moment-dzhozefa-kennedi-ili-istoriya-o-chistilshhike-obuvi.html

https://zen.yandex.ru/media/id/5b0f7d30380d8fbc708f1340/vymiraiuscie-professii-chistilscik-obuvi-5d064256c8a6920d90258434

https://aif.by/vybor/moda/tufelki_s_bleskom_kuda_podevalis_chistilshchiki_obuvi

https://wisecow.com.ua/zhurnalistika/zhurnalistika-v-kino/newspaper-boys.html

https://www.m24.ru/articles/gazety/23082013/24134

https://arzamas.academy/materials/359

———————
2018-2021,
author of the book:
©Tar Sahno
———————
2020-2021,
project manager:
Yana Berezhetskaya
———————
2020-2021,
edited by:
Anastasia Lavender
———————
2021,
cover designed by:
©Alejandro Baigorri
———————
2021,
illustrated by:
©Aleksandra Gammer
———————
Special thanks for the help in the creation of the book:
- Anatolii Uvarov
- Mikalai Kind
- Kristina Sallum
- Eva Vysotska
- Kirill Kaftanik
- Evgeniia Manucharova
- Grant Igitkhanyan
- Olessya Bondar
- Ekaterina Altynnikova
- Roza March
- Ishkhan Badalyan

Made in the USA
Columbia, SC
15 October 2022

69493163R00139